Living Math

*Seeing mathematics in every day life
(and appreciating it more too).*

by D. James Benton

Copyright © 2016 by D. James Benton, all rights reserved.
2020 - some figures and equations enlarged; two errors fixed

Foreword

A friend once told me that he couldn't remember the last time he had used calculus. How sad. I can't remember the last time I didn't use calculus. Mathematics isn't just esoteric busywork. It's the language of cause and effect–a description of why things work the way they do. Mathematics connects the dots in the picture that is the world around us. Understanding the mathematical relationships between objects and events will give you a whole different perspective, like seeing color in a world where others see only black and white. Join me on this adventure, in which I will take you on a tour of *Living Math*.

All of the examples contained in this book,
(as well as a lot of free programs) are available at...
http://www.dudleybenton.altervista.org/software/index.html

Table of Contents

	page
Foreword	i
Chapter 1. Drag Racing	1
Chapter 2: Automobile Drag & Resistance	7
Chapter 3. Vehicle Braking	12
Chapter 4. The Quickest Path	16
Chapter 5. Escape Velocity	21
Chapter 6. Marine Propellers	24
Chapter 7. Axial Fans	37
Chapter 8. Rogue Waves	42
Chapter 9. The Heat Index	46
Chapter 10. Wind Chill	50
Chapter 11. Heating, Cooling, and Entropy	52
Chapter 12. The Perfect Power Plant	55
Chapter 13. The Problem with Giants	58
Chapter 14. Chaos	63
Chapter 15. Probability & Chance	67
Chapter 16. Extrapolating into the Future	72
Chapter 17. Monte Carlo Simulations	78
Chapter 18. Time Series Simulation	86
Chapter 19. Solar Power	90
Chapter 20. Punkin' Chunkin'	96
Chapter 21. Bungee Cords	106
Appendix A. Engine Clearance Simulation Program	109
Appendix B. Projectile Trajectory Using 4th Order Runge-Kutta	112
Appendix C. Punkin' Trajectory and Propulsion	115

Chapter 1. Drag Racing

Drag racing is all about acceleration–getting from here to there as fast as possible. Everyone who has ever driven a car knows that it's harder to accelerate when you're already going 60 mph than from 30 mph, but why is that? Anyone interested in drag racing knows that even doubling the horsepower to weight ratio won't cut the quarter-mile time in half or double the trap speed, but why not and just how much power would it take to accomplish this? Consider the following graph showing the reported performance of 106 cars from several years of a popular car magazine:

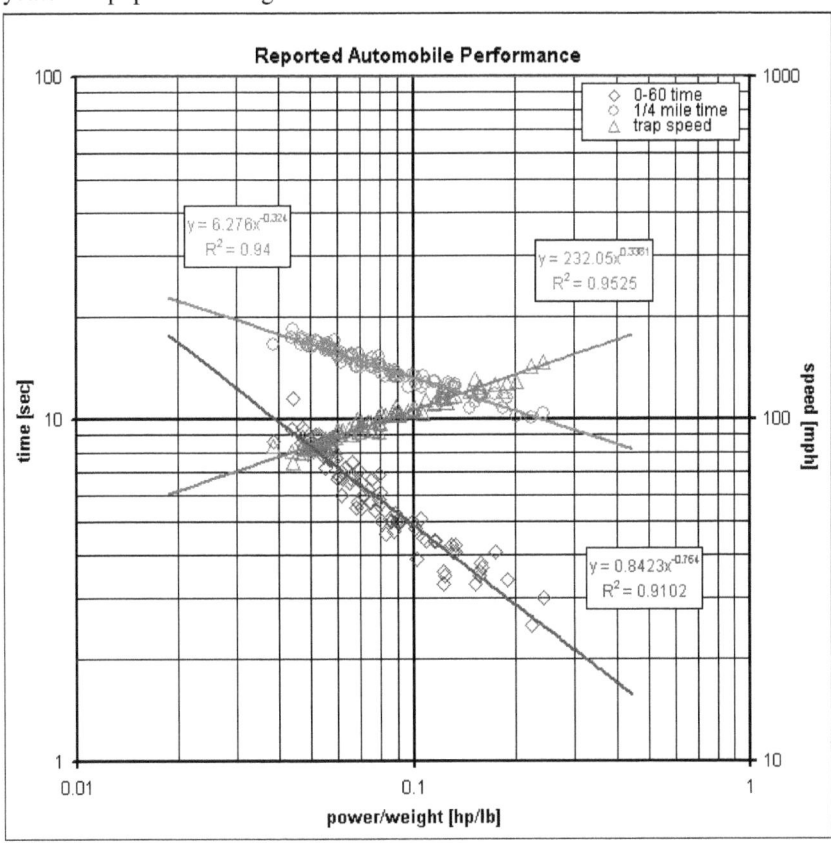

Based entirely on this empirical data, we see that elapsed time in the quarter-mile is inversely proportional to the one-third power of the power-to-weight ratio, trap speed is directly proportional to the one-third power of the power-to-weight ratio, and the 0-60 time is inversely proportional to the three-fourths power of the power-to-weight ratio. Let's now explore the math.

Acceleration is the time rate of change of velocity:

1

$$A = \frac{dV}{dt} \tag{1.1}$$

Velocity is the time rate of change of position:

$$V = \frac{dX}{dt} \tag{1.2}$$

By the chain rule of calculus (combine Equations 1.1 and 1.2, eliminating *dt*):

$$A = V \frac{dV}{dX} \tag{1.3}$$

Equation 1.3 can also be expressed:

$$A = \frac{1}{2} \frac{dV^2}{dX} \tag{1.4}$$

The V^2 in Equation 1.4 explains why it's harder to accelerate when you're already going 60 mph. Force is equal to mass times acceleration:

$$F = mA = \frac{m}{2} \frac{dV^2}{dX} \tag{1.5}$$

Work is the transfer of energy by virtue of a force and is equal to the integral of force with respect to distance:

$$W = \int_0^X F dX = \int_0^X mA dX = \int_0^V \frac{m}{2} dV^2 = \frac{mV^2}{2} \tag{1.6}$$

Equation 1.6 is why kinetic energy equals $mV^2/2$. This important relationship between speed and energy follows directly from calculus. Power is the rate of doing work, or force times velocity:

$$P = \frac{dW}{dt} = FV = mV^2 \frac{dV}{dX} = \frac{m}{3} \frac{dV^3}{dX} \tag{1.7}$$

The V^3 in Equation 1.7 explains why it takes so much more power to accelerate when you're already going 60 mph.

$$PX = \int_0^X P dX = \int_0^V \frac{m}{3} dV^3 = \frac{mV^3}{3} \tag{1.8}$$

Equation 1.8 can be solved for *V*:

$$V = \left[\frac{3PX}{m}\right]^{\frac{1}{3}} \quad (1.9)$$

The green data points and line in the preceding figure are the same as Equation 1.9 except for unit conversions. Equation 1.9 can be solved for power. Given that a quarter-mile is 1320 feet and incorporating unit conversions for mass in pounds[1] and velocity in mph, yields:

$$hp = pounds\left(\frac{mph}{281.1}\right)^3 \quad (1.10)$$

Combining Equations 1.1, 1.5, and 1.7 yields:

$$Pt = \int_0^t P dt = \int_0^t mV\left(\frac{dV}{dt}\right) dt = \int_0^V \frac{m}{2} dV^2 = \frac{mV^2}{2} \quad (1.11)$$

Combining Equations 1.9 and 1.11 yields:

$$t = \left[\frac{9mX^2}{8P}\right]^{\frac{1}{3}} \quad (1.12)$$

Equation 1.12 is identical to the red data points and line except for unit conversions. This equation can also be solved for power. Incorporating the same unit conversions yields:

$$hp = \frac{110.8\, pounds}{\sec^3} \quad (1.13)$$

Equations 10.1 and 10.13 are an indication of the average power delivered to the wheels. This next figure is a comparison of these two equations to the reported values peak engine power. As indicated by the green points, trend line, and equation (y=0.55x), based on the quarter-mile speeds, the average power delivered is only about 55% of the reported peak power. As indicated by the red points, trend line and equation (y=0.47x), based on the quarter mile times, the average power delivered is only about 47% of the reported peak power.

Equation 1.11 can be solved for time:

$$t = \frac{mV^2}{2P} \quad (1.14)$$

Equation 1.14 doesn't match the blue data in the figure for 0-60 mph. Up to this point we have assumed the power to be constant, but it isn't and that's why

---

[1] Note that mass and weight are the same in pounds at standard gravity (32.174 ft/sec^2).

Equation 1.14 differs from the data by a power of one-fourth. The variation in over time is more evident in the 0-60 time than in the quarter-mile time, because the engine is just getting up to full power during this shorter interval of time. We can use this experimental information to infer the typical engine response and compare this to dynamometer data.

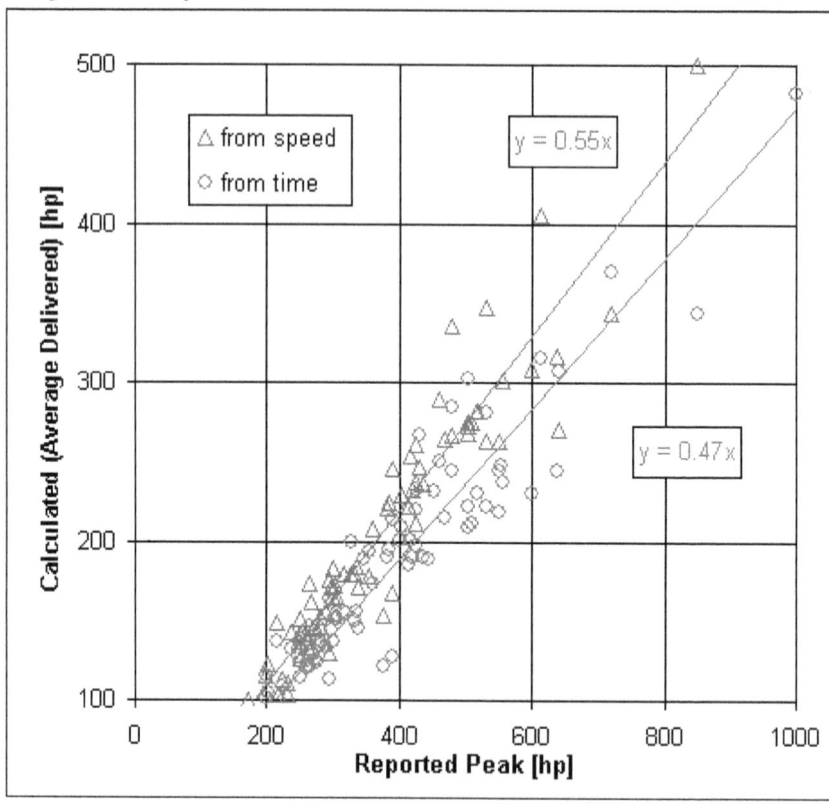

If the 0-60 time is inversely proportional to the power-to-weight ratio raised to an exponent of approximately 3/4ths, then the power must be proportional to time raised to an exponent of approximately 4/3rds. Since the velocity increases with the integral of power with respect to time (Equation 1.11), the power must increase proportional to the time raised to the 1/3rd power, as illustrated by the following integral:

$$\int x^{\frac{1}{3}} dx = \frac{3}{4} x^{\frac{4}{3}} \tag{1.15}$$

We can use the exact coefficients from the regression rather than the estimated thirds and fourths.

$$t_{60} = 0.8423105 \left(\frac{m}{P}\right)^{0.763982} \quad (1.16)$$

The power, P, must then be equal to some constant (for a particular vehicle) times time raised to the power 1/0.763982 or 1.308931362.

$$P = \alpha \, t^{1.308931362} \quad (1.17)$$

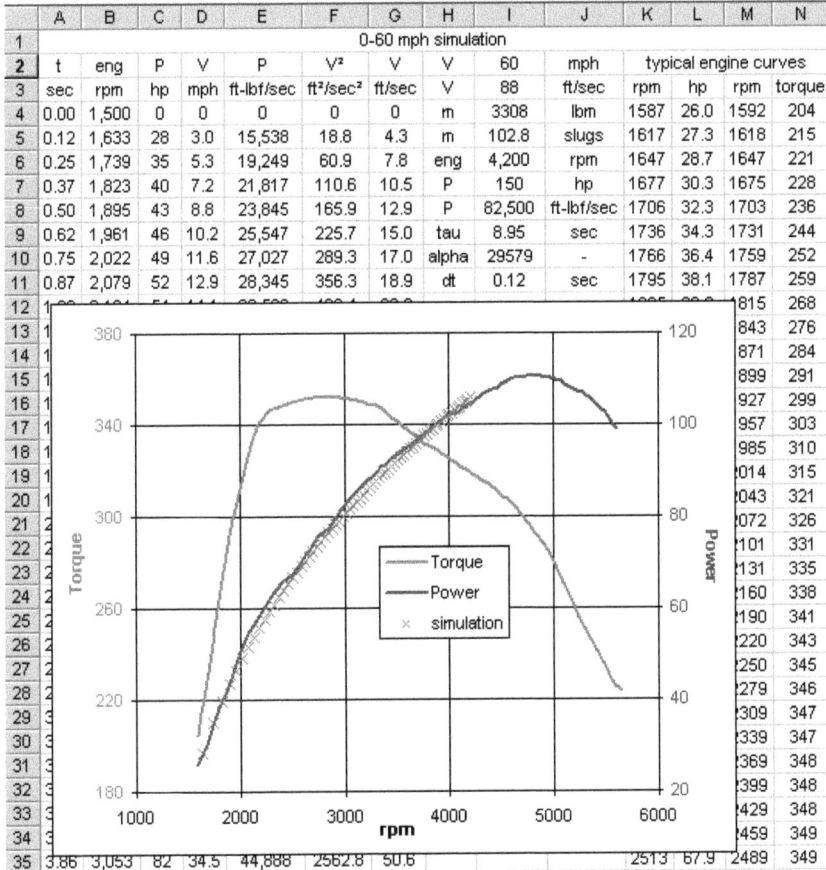

Furthermore, the engine speed in the range of 0 to 60 mph is roughly proportional to the velocity. The engine doesn't stop rotating when the vehicle is stationary, due to the clutch or torque converter. A typical staging speed would be perhaps 1500 rpm. We can put all of this into a spreadsheet and calculate the power, ground speed, and engine rotational speed. A simple trapezoidal formula can be used to integrate the equations, as follows:

$$y(t + \Delta t) = y(t) + \Delta t \left[\frac{dy}{dt}\right]_{avg} \qquad (1.18)$$

You might be wondering... Can you really little buy a box with a microcomputer and accelerometer in it, stick it on the dashboard, drive to the truck stop and get your car weighed, punch in the number, go out on the highway, floor it, a get a reasonably accurate measure of horsepower delivered to the wheels? The answer is...

Yes, because calculus is consistent with experience!

Chapter 2: Automobile Drag & Resistance

In the previous chapter we implicitly accounted for drag and resistance because we used actual times and speeds. In this chapter we will see how to account for these things independently. The following information is quite dated[2], but still useful in estimating *form drag* or wind resistance on cars of various shapes, as the air hasn't changed. This figure identifies categories of shape factors:

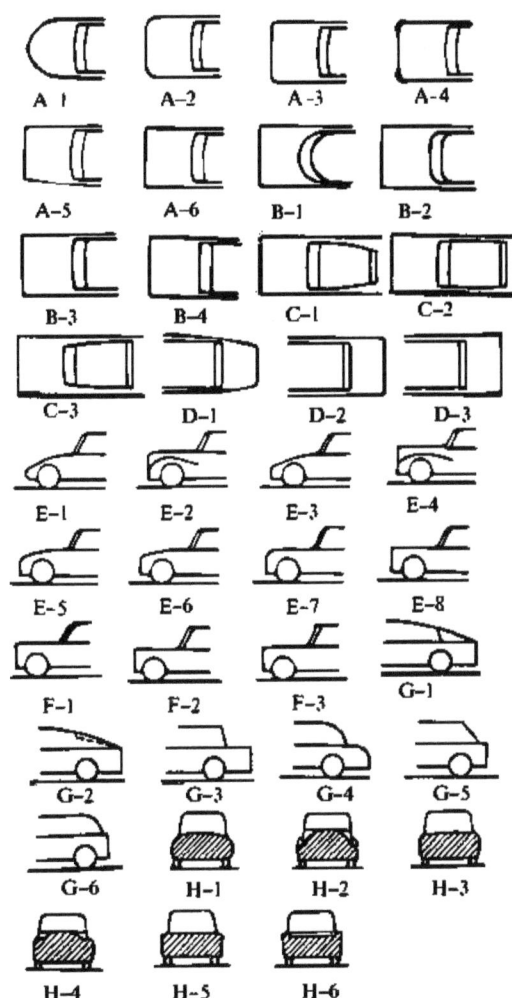

---

[2] "A Method of Estimating Automobile Drag Coefficients," R. G. S. White, Society of Automotive Engineers Paper No. 690189, 1969.

The categories above are used with the following table of values:
Include one rating from each of the nine categories listed (A through I).
A. Plan view, front end
  A-1 Approximately semicircular (add 1)
  A-2 Well-rounded outer quarters (add 2)
  A-3 Rounded corners without protuberances (add 3)
  A-4 Rounded corners with protuberances (add 4)
  A-5 Squared tapering-in corners (add 5)
  A-6 Squared constant-width front (add 6)
B. Plan view, windshield
  B-1 Full wrap-round approximately semicircular) (add 1)
  B-2 Wrapped-round ends (add 2)
  B-3 Bowed (add 3)
  B-4 Flat (add 4)
  Add 1 for upright windshield; add 1 for prominent flanges or rain gutters.
C. Plan view, roof
  C-1 Well- or medium tapered to rear (add 1)
  C-2 Tapering to front and rear (max. width at BC post) or approximately constant width (add 2)
  C-3 Tapering to front (max. width at rear) (add 3)
D. Plan view, lower rear end
  D-1 Well- or medium tapered to rear (add 1)
  D-2 Small taper to rear or constant width (add 2)
  D-3 Outward taper (or flared-out fins) (add 3)
E. Side elevation, front end
  E-1 Low, rounded front, sloping up (add 1)
  E-2 High, tapered, rounded hood (add 1)
  E-3 Low, squared front, sloping up (add 2)
  E-4 High, tapered, squared hood (add 2)
  E-5 Medium-height, rounded front, sloping up (add 3)
  E-6 Medium-height, squared front, sloping up (add 4)
  E-7 High, rounded front, with horizontal hood (add 4)
  E-8 High, squared front, with horizontal hood (add 5)
  Add 3 for separate fenders; add 4 for open front to fenders (above bumper level); add 2 for raised built-in headlamps; add 4 for small separate headlamps; add 7 for large separate headlamps
F. Side elevation, windshield peak
  F-1 Rounded (add 1)
  F-2 Squared (including flanges or gutters) (add 2)
  F-3 Forward-projecting peak (add 3)
G. Side elevation, rear roof/trunk
  G-1 Fastback (roof line continuous to tail) (add 1)
  G-2 Semi-fastback (with discontinuity in line to tail) (add 2)
  G-3 Squared roof with trunk rear-edge squared (add 3)
  G-4 Rounded roof with rounded trunk (add 4)
  G-5 Squared roof with short or no trunk (add 4)
  G-6 Rounded roof with short or no trunk (add 5)

Add 3 for high fins or sharp, longitudinal edges to trunk; add 2 for separate fenders
Note: the trunk is assumed to be rounded laterally.
H. Front elevation, cowl and fender cross-section at windshield
   H-1 Flush hood and fenders, well-rounded body sides (add 1)
   H-2 High cowl, low fenders (add 2)
   H-3 Hood flush with rounded-top fenders (add 3)
   H-4 High cowl, with rounded-top fenders (add 3)
   H-5 Hood flush with square-edged fenders (add 4)
   H-6 Depressed hood, with high square-edged fenders (add 5)
   Fender mirrors: include in protuberances if at the fender leading end; otherwise add 1.
I. Underbody Drag Rating
   I-1 For integral, flush floor, little projecting mechanism. Examples: Porsche coupe, Citroen DS19, Saab 96 (add 1)
   I-2 Intermediate between descriptions for (1) and (3). Examples: Volkswagen 1300 (add 2)
   I-3 Integral, projecting structure and mechanism. Examples: Mercedes 300 SE, Ford Falcon and Mustang (add 3)
   I-4 Intermediate between descriptions for (3) and (5). Examples: Oldsmobile Toronado, Jenson 541, Ferrari 300 GTB (add 4)
   I-5 Deep chassis. Examples: Typical U.S. full size automobiles with separate body and frame construction (add 5)

These shape factors are used in the following equation:

$$C_D = 0.16 + 0.0095 \sum factors \qquad (2.1)$$

The form drag or wind resistance is calculated from this next equation:

$$F_D = \frac{C_D A_P \rho V^2}{2} \qquad (2.2)$$

In Equation 2.2, A_P is the projected area (as presented to the wind) and ρ is the density of air (approximately 0.0765 lbm/ft^3 or 0.001225 g/cm^3). The following figure illustrates projected area:

In this case the projected area is 85.3% of the overall height times width. This profile is similar to the Audi A7, in which case A=1, B=1, C=1, D=1, E=1, F=1, G=1, H=1, and I=1. C_D would then be 0.16+0.0095*9=0.246. The dimensions (HWL) for this vehicle are approximately 56x75x196 inches, making the projected frontal area, A_P, approximately 24.9 ft². The following table and graph shows the results of this calculation:

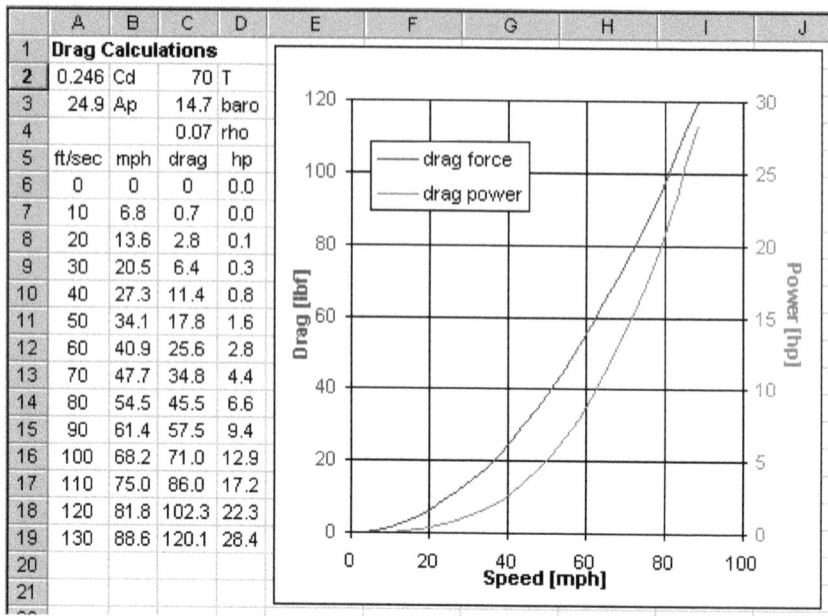

At 88.6 mph the calculated drag force is 120.1 lbf, which requires 28.4 hp delivered to the wheels to overcome it. We must also consider rolling resistance. The coefficient of rolling resistance, C_R, for a car with typical tires on cement or asphalt varies from about 0.010 to 0.015. Special, very low resistance, tires may be as low as 0.005. Large truck tires vary from about 0.0045 to 0.008. The rolling force is calculated by the following equation:

$$F_R = C_R W \qquad (2.3)$$

In this case, **W**, is the vehicle weight. Taking the lower value for the coefficient of rolling resistance and a weight of 4300 lb, we get the following:

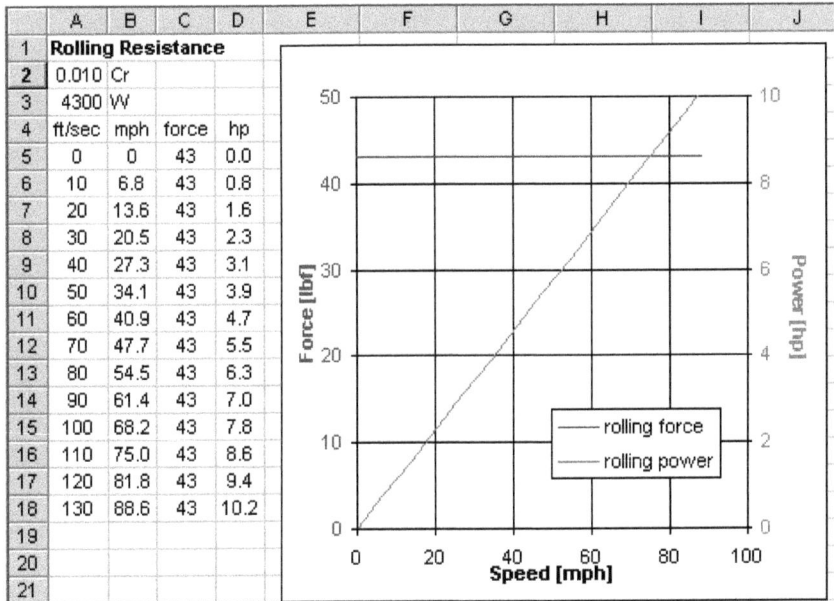

It takes another 10.2 hp to overcome the rolling resistance, making the total 38.6 hp at 88.6 mph–and this is for a streamlined sports car. At this speed and with the reported gear ratio, the engine would be turning approximately 4000 rpm. At this rotational speed, this power would require almost 52 ft-lbf of torque. Is it any wonder so much fuel is consumed each year by motor vehicles?

Chapter 3. Vehicle Braking

For every automotive start, there is a stop–hopefully by brakes and not a tree. The coefficient of friction, *f*, is defined as the force divided by the weight. As weight is equal to the mass times the acceleration of gravity and force is equal to the mass times acceleration, the coefficient of friction is equal to the acceleration divided by gravity, or:

$$f = \frac{|a|}{g} \qquad (3.1)$$

Equation 3.1 also illustrates why it is common to report acceleration as so many "g's". An excellent source of data available on the Web is, "Drag Factor and Coefficient of Friction in Traffic Accident Reconstruction," by Lynn B. Fricke and J. Stannard Baker, published in 1990 by the Northwestern University Traffic Institute, and appearing as Topic 862 in the *Traffic Accident Investigation Manual*. The following table appears in this document:

Coefficients of Friction for Various Roadway Surfaces*

Description of Surface	Dry <30 mph from	Dry <30 mph to	Dry >30 mph from	Dry >30 mph to	Wet <30 mph from	Wet <30 mph to	Wet >30 mph from	Wet >30 mph to
Portland Cement								
New, Sharp	0.80	1.20	0.70	1.00	0.50	0.80	0.40	0.75
Traveled	0.60	0.80	0.60	0.75	0.45	0.70	0.45	0.65
Traffic Polished	0.55	0.75	0.50	0.65	0.45	0.65	0.45	0.60
Asphalt or Tar								
New, Sharp	0.80	1.20	0.65	1.00	0.50	0.80	0.45	0.75
Traveled	0.60	0.80	0.55	0.70	0.45	0.70	0.40	0.65
Traffic Polished	0.55	0.75	0.45	0.65	0.45	0.65	0.40	0.60
Excess Tar	0.50	0.60	0.35	0.60	0.30	0.60	0.25	0.55
Gravel								
Packed, Oiled	0.55	0.85	0.50	0.80	0.40	0.80	0.40	0.60
Loose	0.40	0.70	0.40	0.70	0.45	0.75	0.45	0.75
Cinders								
Packed	0.50	0.70	0.50	0.70	0.65	0.75	0.65	0.75
Rock								
Crushed	0.55	0.75	0.55	0.75	0.55	0.75	0.55	0.75
Ice								
Smooth	0.10	0.25	0.07	0.20	0.05	0.10	0.05	0.10
Snow								
Packed	0.30	0.55	0.35	0.55	0.30	0.60	0.30	0.60
Loose	0.10	0.25	0.10	0.20	0.30	0.60	0.30	0.60

*Not applicable for large, heavy trucks.

As expected, the coefficient of friction varies vehicle speed, weather, and surface type. No doubt, tire design and condition as well as weight distribution and control also influence this important measure of breaking performance. The variation of f with speed is shown in this next figure, which also appears in this document.

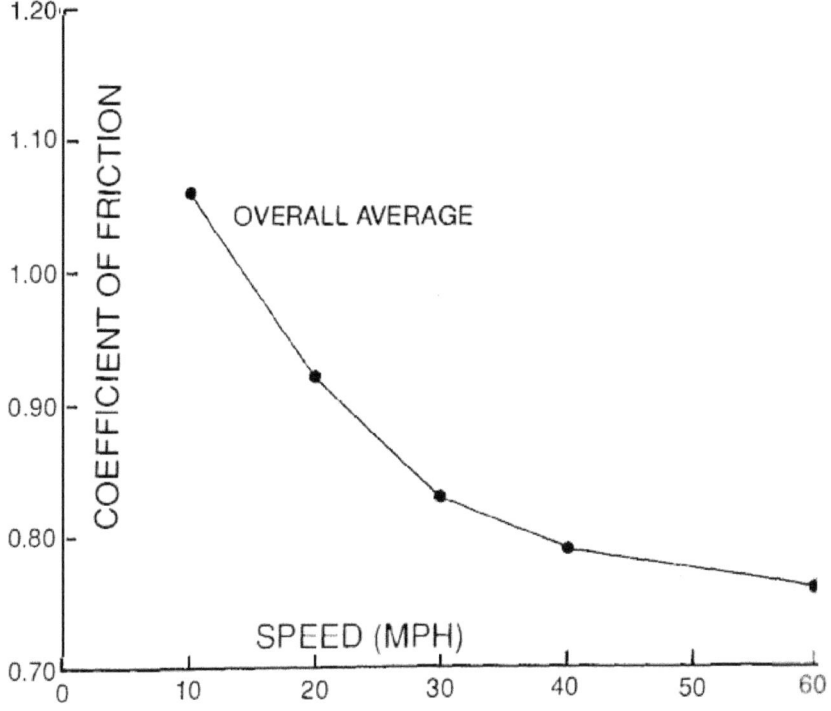

As expected, f diminishes with increasing speed. This curve may be approximated up to 60 mph by the following formula:

$$f = \frac{4.85606}{\left(\dfrac{mph}{60}+0.406884\right)\left[\left(\dfrac{mph}{60}\right)^2 - 5.23051\left(\dfrac{mph}{60}\right)+8.77546\right]} \quad (3.2)$$

The coefficient of friction also depends on whether the tires are slipping or skidding. This next figure is also from Fricke and Baker:

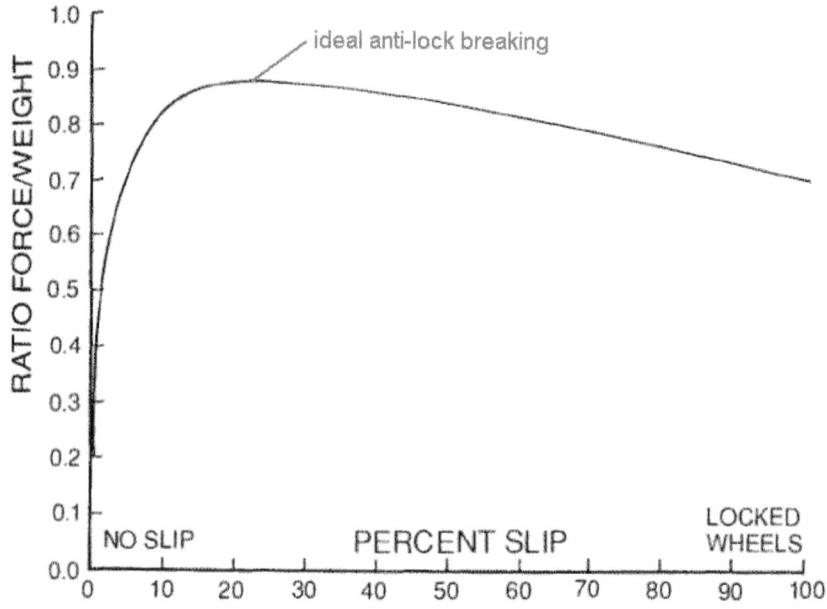

The maximum occurs at about 20% slip, which is the ideal for anti-lock braking. This curve may be approximated by the following formula:

$$f = 0.108988 + (3.98684 + 4.99510s)\sqrt{s} - (7.04270 + 1.34078s)s \tag{3.3}$$

In this case the slip, s, varies from 0 to 1, not 0 to 100. Combining Equation 1.3 and 3.1 we get:

$$fg = -V\frac{dV}{dx} \tag{3.4}$$

Separating and integrating yields:

$$\int_0^D dx = \int_0^V \frac{VdV}{fg} \tag{3.5}$$

As f may be a function of V, it must be inside the integral on the right. Swapping the limits 0 to V eliminates the minus sign. If f is constant, this reduces to:

$$D = \frac{V^2}{2fg} \tag{3.6}$$

14

If f varies with V, as in the previous figure, this expression can easily be integrated, as in the following spreadsheet:

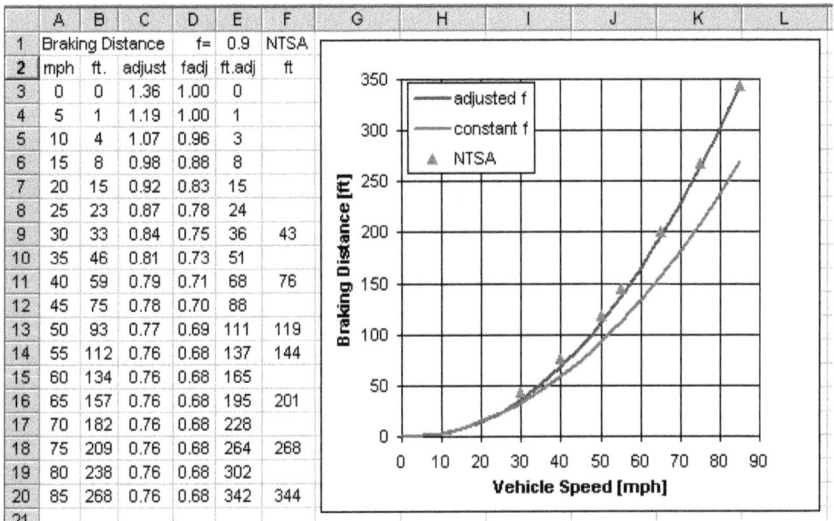

The green curve (constant f) is directly proportional to V^2; whereas, the blue curve (variable f) is approximately proportional to $V^{2.15}$.

Chapter 4. The Quickest Path

Whether it's skiing, snowboarding, or skateboarding, the quickest path between two points can be more important than the shortest distance. Most everyone knows the latter is a straight line, but what about the former? Mathematicians have considered this problem since the time of the ancient Greeks. The problem is illustrated below with marbles:

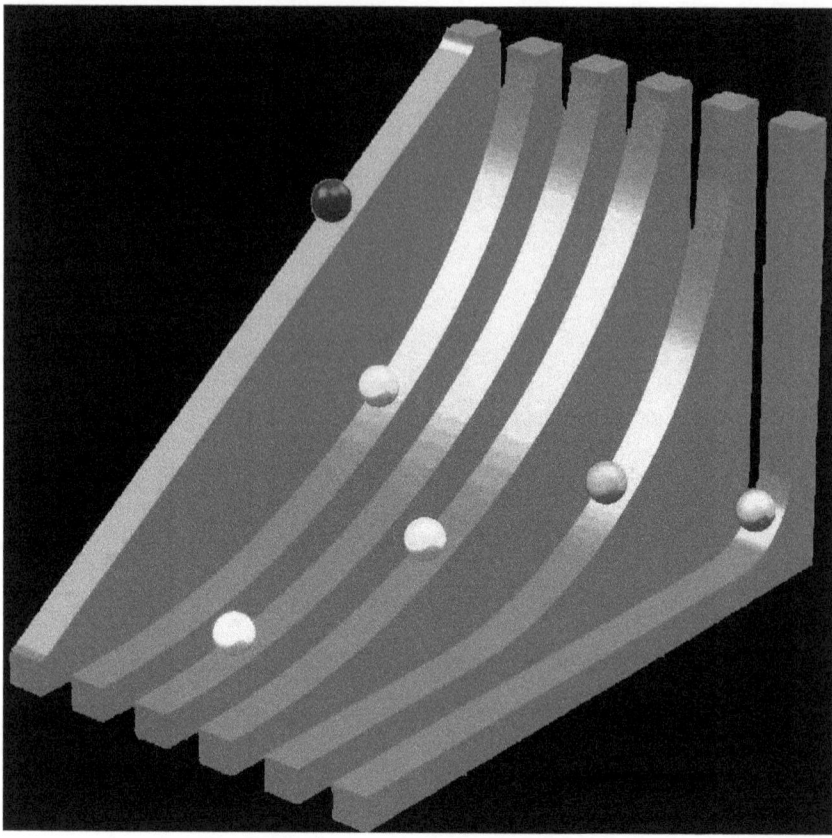

The quickest path is called a brachistochrone and is shown in the following figure, as the arc formed by a point on a rotating circle or cycloid:

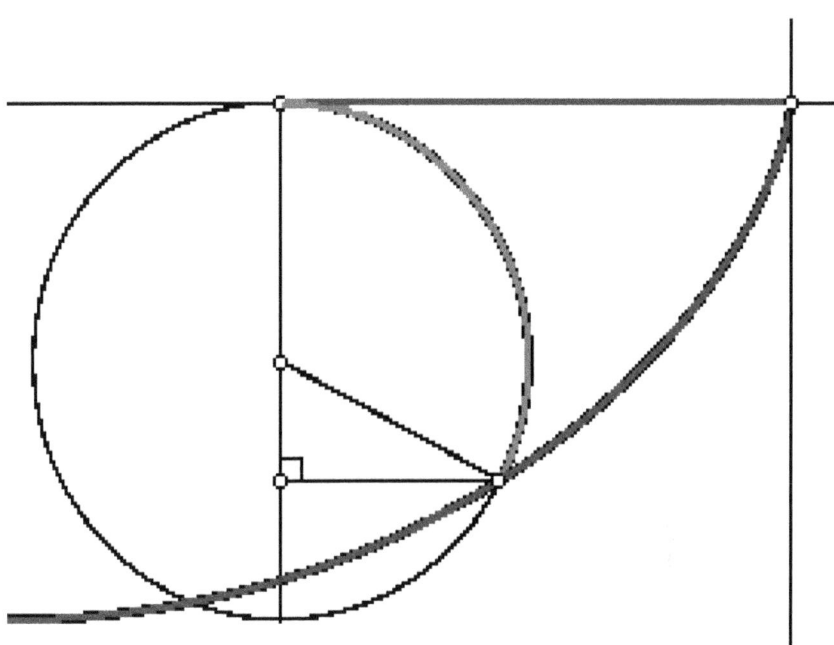

The mathematical proof of this was first presented by Isaac Newton and is a classic problem in the Calculus of Variations. The final speed will be the same, regardless of the path; because the change in kinetic energy is equal to the change in potential energy, that is: $gh=V^2/2$.

This path makes since, because the object begins at rest and must be accelerated. The quickest path results from taking a sharp drop to get the object going as soon as possible. Some amusement park rides are designed with this shape, as illustrated by the *Superman Escape from Krypton®* at Six Flags®:

Now this is math that everyone can appreciate!

Denote the horizontal distance from the starting point as *x* and the vertical *y*. The distance along the path will be called *s*. The differential distance along the path becomes:

$$ds = \sqrt{dx^2 + dy^2} \tag{4.1}$$

From the conservation of energy (neglecting friction and air resistance), we know that:

$$V = \frac{ds}{dt} = \sqrt{2gy} \qquad (4.2)$$

Separating and integrating yields:

$$t = \int dt = \int_0^x \frac{\sqrt{1 + \left(\frac{dy}{dx}\right)^2}}{\sqrt{2gy}} \qquad (4.3)$$

We seek $y(x)$ such that this integral is a minimum (i.e., the quickest path). The branch of mathematics that deals with problems of this type is called the calculus of variations. In this we consider problems that can be expressed by the following integral:

$$I = \int F\left(x, y, \frac{dy}{dx}\right) dx \qquad (4.4)$$

In this case, I is what we want to minimize, F is the unknown function, x and y are the independent and dependent variables. The stationary value (i.e., minimum or maximum) will satisfy the following partial differential equation:

$$\frac{\partial F}{\partial y} = \frac{d}{dx}\left(\frac{\partial F}{\partial \left(\frac{dy}{dx}\right)}\right) \qquad (4.5)$$

This is known as the Euler-Lagrange equation. Substituting Equation 2.3 into 2.5 and rearranging yields:

$$\frac{\partial F}{\partial \left(\frac{dy}{dx}\right)} = \frac{\left(\frac{dy}{dx}\right)}{\sqrt{2gy}\sqrt{1 + \left(\frac{dy}{dx}\right)^2}} \qquad (4.6)$$

The first step in solving this partial differential equation yields:

$$\sqrt{2gy}\sqrt{1 + \left(\frac{dy}{dx}\right)^2} = c \qquad (4.7)$$

Here, c is a constant. Squaring both sides and rearranging yields:

$$y\left[1+\left(\frac{dy}{dx}\right)^2\right] = \frac{1}{2gc^2} = k^2 \qquad (4.8)$$

For convenience, k^2 has been substituted for *1/2gc²*. Equation 2.8 has the following solution:

$$x = \frac{k^2(\theta - \sin\theta)}{2} \qquad (4.9)$$

$$y = \frac{k^2(1-\cos\theta)}{2} \qquad (4.10)$$

The factor **2** could have been included in k^2, the square isn't essential, and some other variable could have been used instead of θ. These things have been selected so that Equations 2.9 and 2.10 will have the same form as the *cycloid*:

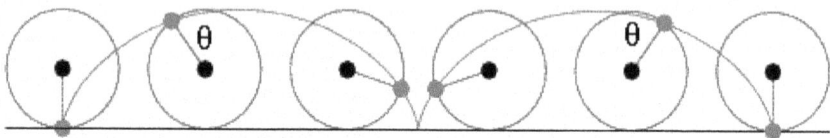

So the quickest path turns out to be a geometrically simple shape. The ancients figured this out, but couldn't prove it mathematically until the 17th century.

Chapter 5. Escape Velocity

There's always someone suggesting we put stuff up into space with a *magnetic elevator* or a winch connected to a tethered satellite (perhaps the Moon?). These self-proclaimed geniuses imply that the losers at NASA are either idiots or in cahoots with rocket salesmen. Duh! If they're so smart, why haven't they thought of this?

The ancient Greeks had a concept of physical harmony that we find humorous today, but one example is worth relating here. To the ancient Greek mind, rocks fall to the ground when you let go of them, because they *like* being in low places. By this same reasoning, meteorites really prefer Earth to space, because they fall with great enthusiasm. Aside from the humor, there is an underlying truth in this. Putting stuff up into space is making it do something it doesn't want to do–or at least doesn't naturally do on its own.

Let's begin this inquiry with the derivation of escape velocity. Newton's law of gravitation is:

$$F = \frac{GmM}{R^2} \quad (5.1)$$

By the time Newton presented his equation (and began his argument with Hooke over who came up with it first), Kepler had already supplied the proof, based on the work of Copernicus and Tyco Brahe. F is the gravitational force between two bodies, G is a universal constant, m is the mass of the object, M is the mass of the Earth, and R is the distance between the centers of mass. Since we already know that on the surface of the Earth, $F=mg$, then $g=GM/R^2$. We can measure g and the radius of the Earth. As soon as scientists were able to measure G, they could calculate the mass of the Earth.

The reason the Moon doesn't escape from the Earth is because it's stuck in a potential energy well. You would have to increase its energy in order to get it out of the well and away from the Earth. Potential energy is the integral of the gravitational force with respect to the distance. At escape velocity, the kinetic energy is equal to the potential energy deficit so that when R=∞ both the kinetic and potential energy are zero and the object is free and at rest. This condition is expressed by the following equation:

$$\int_0^{V_e} mV dV = \int_R^{\infty} \frac{GmM}{r^2} dr \quad (5.2)$$

Integrating and substituting $g=GM/R^2$ into the result yields:

$$V_e = \sqrt{2gR} \quad (5.3)$$

The result is 24,688 mph or 11.037 km/s. Recall that the kinetic energy was equal to the potential energy deficit, or:

$$KE = \frac{mV_e^2}{2} = -PE = gRm \qquad (5.4)$$

To put an object into space (i.e., beyond Earth orbit), you must increase its energy by *gRm*. For a one-pound object this would be 20,375,520 ft-lbf or 10.3 horsepower-hours. If a horse could drag a one-pound object into space, it would take 10.3 hours to do so. This energy is equivalent to 0.228 gallons or 1.41 pounds of gasoline. It takes at least 41% more fuel than the object you're trying to lift, even ignoring every practical detail.

A pound of gasoline can't even propel itself into orbit.
For space travel, you'll need something with a little more kick.

Factor in that a rocket must lift off with all of the fuel plus the engines are not 100% efficient plus you'll need the oxygen to burn the fuel if you plan to leave the Earth's atmosphere behind, and you get a payload ratio of approximately 1:1600. The mass of the Apollo module was 1840 kg and the mass of the Saturn V rocket that put it up there was 3,038,500 kg! Reported payload ratios on the order of 1:25 include the weight of the rocket and all the stuff that's shed and falls back to Earth, so they're not a fair comparison.

Let's evaluate the magnetic elevator concept. You can get floating magnets like the following at any toy store:

Like gravity, the force exerted by the magnets diminishes with the square of the distance. The problem with a magnetic elevator is not how you're going to push the stuff up; it's how you're going to push the magnets up. A magnetic elevator only works for a few inches, and then you have to raise the magnets. If you must raise the magnets, what benefit is there in a magnetic elevator? If your magnetic elevator were to climb up a shaft or cable, what's the difference between that and a winch? You could run a motor to raise the winch and a generator to lower it, but you'd lose power both ways. It's not going to be free.

You have to put gas in a Prius. It's not free transportation.

The idea of space tethering has been around for quite a while. Set aside for the moment, what you're going to tie this hypothetical rope to. Nik Wallenda set the world's record when he walked a 1,560-foot-long wire at the Wisconsin State Fair in 2015. Now that's impressive... but not if you're headed for space. The problem with this idea is that the cable couldn't support it's own weight, let alone that of the payload. If you find someone foolish enough to pay you to try this, make sure you have an airtight disclaimer for cross wind damage. If you're still not convinced I highly recommend the following two articles:

http://what-if.xkcd.com/58/

http://www.quora.com/If-a-cable-could-be-attached-to-the-Moon-and-left-to-hang-within-a-mile-of-the-Earth-could-it-be-used-as-a-space-elevator

Chapter 6. Marine Propellers

Propellers are fascinating. There are countless designs and, like buildings, each reflects the objectives of the architect. The shape of a propeller reveals its intended application. Large sweeping blades are used when thrust is of primary importance and small slicing blades are for speed. Four or more blades are preferable for pulling and most racing props have only two blades.

The two most important dimensions of a propeller are pitch and diameter. Pitch is the distance traveled in one revolution if there were no slippage. The diameter is larger than the pitch for thrusting props and the pitch is larger than the diameter for racing props. One of my most prized possessions is a two-blade Oakland-Johnson prop with a 9½-inch diameter and 32-inch pitch, similar to the one pictured below:

The shape above is typical of racing props and the one below is for a tugboat:

There are other objectives besides thrust and speed, such as running through weeds without being entangled, which is the purpose of the following prop:

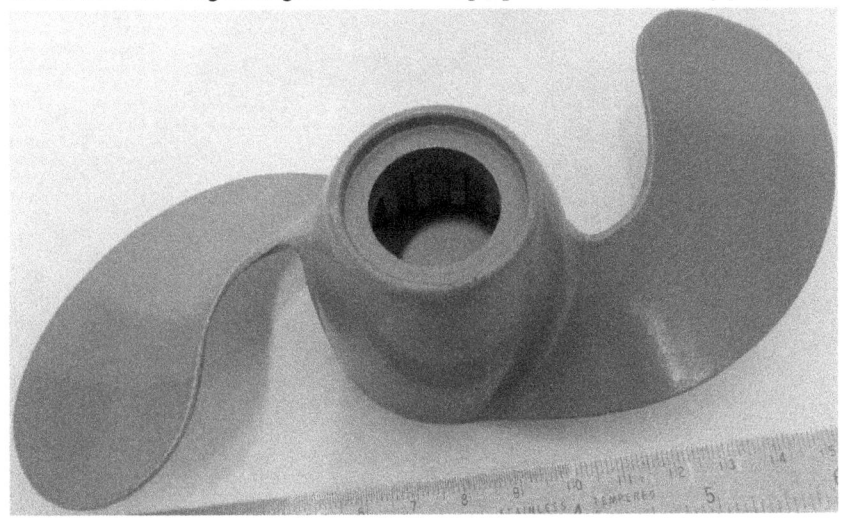

25

Some hull designs and engine placement result in stern squat. When stern lift is important, a cleaver is used. In this design the trailing edge of the blades ends abruptly, as illustrated in the next figure:

The increased stern lift of this design comes at a price, which is less control over the thrust cone, often called *hole shot*. This next figure shows an essentially constant thrust cone. Even the through-hub exhaust flows straight back.

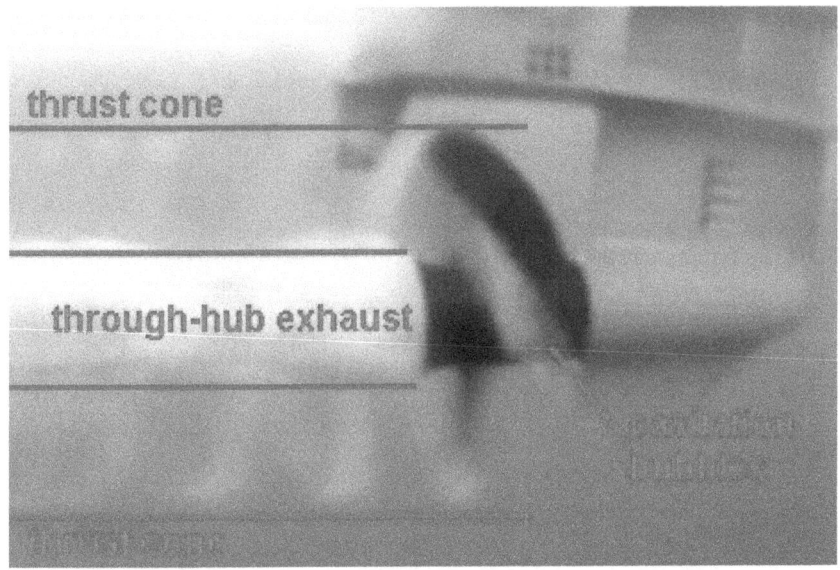

Some cavitation can be seen in the faint spiral pattern of bubbles, forming at the blades and trailing away to the left. A divergent cone results in more thrust and a convergent cone results in more speed. The following figure illustrates this difference:

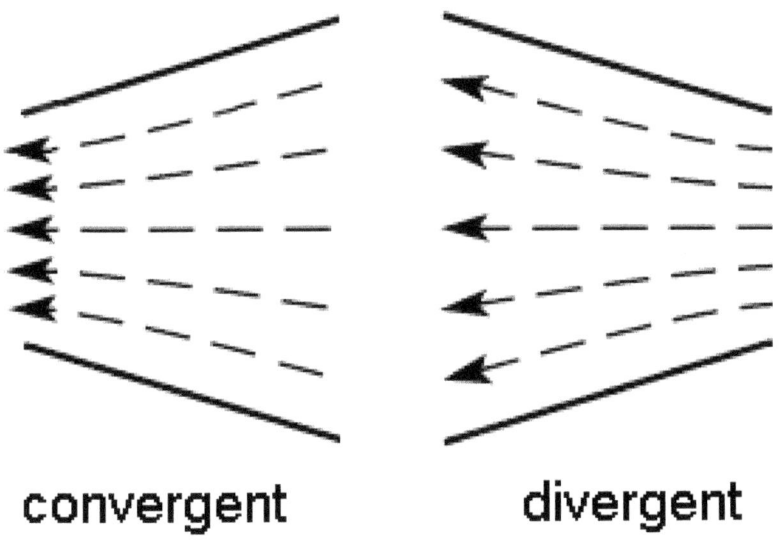

27

If thrust is the goal (i.e., a ski boat or tugboat), a divergent flow moves more water slowly. If speed is the goal, a convergent flow moves less water faster. Ultimately, you must push water out the back faster than you hope to go forward. There is also an optimum ratio, as we shall see. Different blade shapes produce convergent and divergent flows.

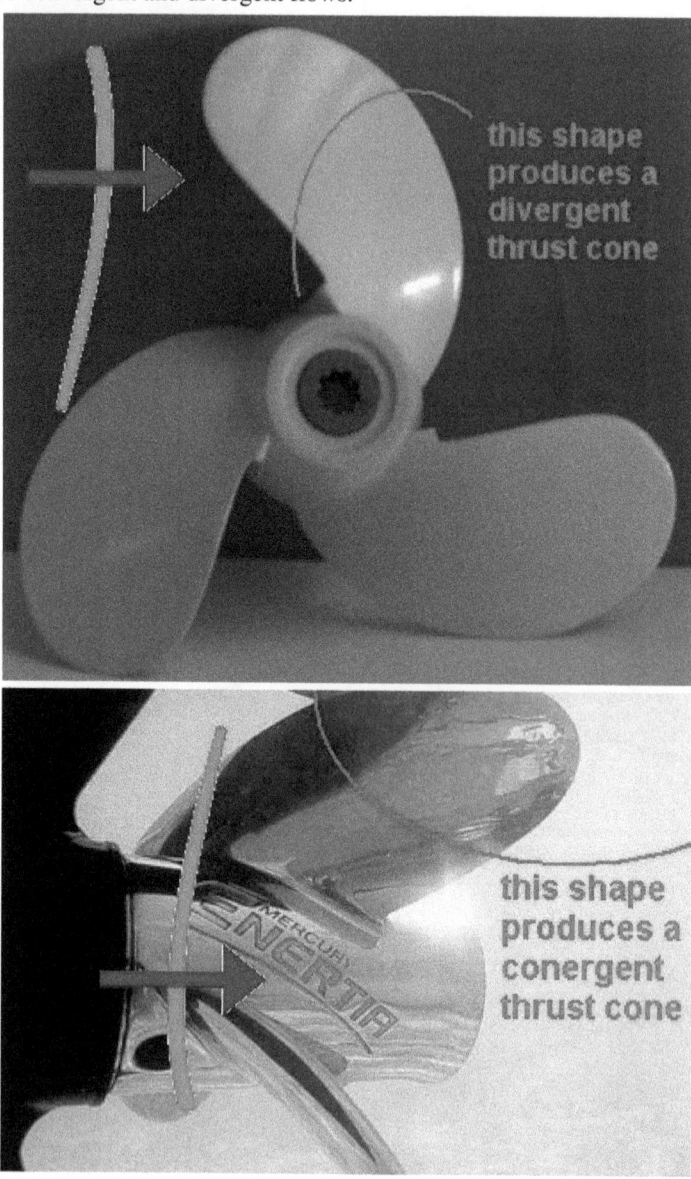

this shape produces a divergent thrust cone

this shape produces a conergent thrust cone

First, we must delve deeper into pitch. The pitch at a point is:
$$\phi = 2\pi r \tan(\theta) \tag{6.1}$$
Here r is the radius at the point and θ is the angle with respect to the shaft. The tangent of 45° is 1, which would make the pitch equal to the radius. If you want to produce a thrust cone that is neither convergent nor divergent, you must vary the angle along the blade so that $r \cdot tan(\theta)$ is a constant. The blades for such a design appear to flatten out away from the hub. This is especially noticeable in aircraft propellers, as in the following figure:

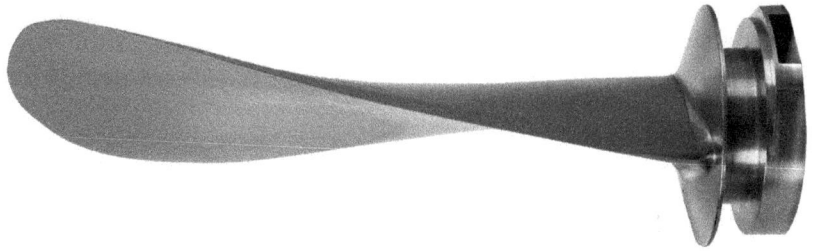

The total (or average) pitch is the integral over the blade surface, or:
$$\Phi = \frac{\iint 2\pi r \tan(\theta)\, dA}{A} \tag{6.2}$$

Thrust results from accelerating the fluid (water or air). The velocity must be faster leaving the prop than the prop is moving through the fluid. Only the increase in velocity along the direction of motion is useful. Swirl and turbulence are just wasted power. This is the inspiration for coupled, contra-rotating props, as illustrated in this next figure:

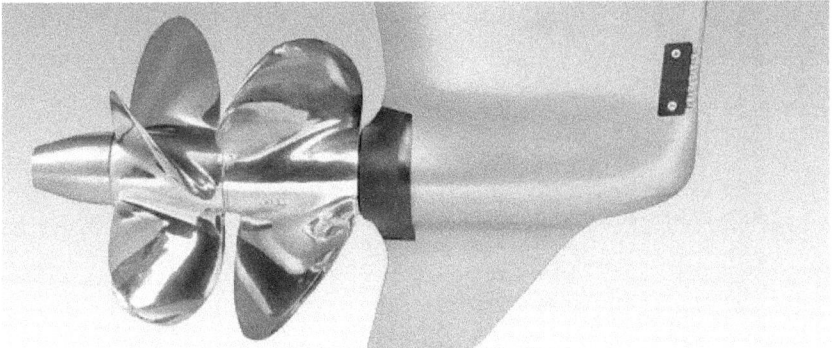

Ideally, the thrust would be given by the following formula, where D_P is the diameter of the prop and D_H is the diameter of the hub:

$$F = \frac{\pi(D_P^2 - D_H^2)}{4}\rho \Delta V(V + \Delta V) \tag{6.3}$$

The first term is the effective area, ρ is the density, V is the velocity of the incoming flow, and ΔV is the increase in velocity provided by the propeller. If there were no slip, the average velocity of the fluid at the propeller would be:

$$V_P = \Phi \Omega \tag{6.4}$$

where Ω is the rotational speed. For example, consider a 115 hp outboard motor turning 5800 rpm. Typical gears in the lower unit would have a tooth count of 12 and 28 for the pinion and prop shaft, respectively, for a gear ratio of 2.33:1. The most common prop for this engine would be a three-blade aluminum having a diameter of 13 inches and a pitch of 21 inches. The hub is about 4⅛ inches. In this case VP would be 72.5 ft/sec or 49.4 mph. Clearly, this engine/prop combination couldn't push a boat faster than 49 mph. The ideal thrust given by Equation 6.3 is listed in the following table along with other calculations:

	A	B	C	D	E	F	G	H	I	J	K
1	typical 115 hp outboard										
2	5800	rpm									
3	2.33	gear ratio									
4		13	diameter [in]								
5		4.16	hub [in]								
6		21	pitch [in]								
7		72.5	Vp [ft/sec]								
8		62.4	density [lb/ft³]								
9	boat	boat	thrust	gross	prop	prop	prop	thrust	gross	net	prop
10	mph	ft/sec	lbf	hp	slip	ft/s	mph	lbf	hp	hp	eff
11	0.0	0	8433	1112	68%	23.3	15.9	872	115	0	0%
12	3.4	5	7852	1035	64%	26.0	17.7	872	115	8	7%
13	6.8	10	7270	958	60%	28.8	19.7	872	115	16	14%
14	10.2	15	6689	882	56%	32.0	21.8	872	115	24	21%
15	13.6	20	6107	805	51%	35.4	24.1	872	115	32	28%
16	17.0	25	5525	728	46%	39.0	26.6	872	115	40	34%
17	20.5	30	4944	652	41%	42.7	29.1	872	115	48	41%
18	23.9	35	4362	575	36%	46.7	31.8	872	115	56	48%
19	27.3	40	3780	498	30%	50.7	34.6	872	115	63	55%
20	30.7	45	3199	422	24%	54.9	37.4	872	115	71	62%
21	34.1	50	2617	345	18%	59.2	40.4	872	115	79	69%
22	37.5	55	2036	268	12%	63.6	43.3	872	115	87	76%
23	40.9	60	1454	192	6%	68.0	46.4	872	115	95	83%
24	**42.5**	62.33	1183	156	3%	70.1	**47.8**	**872**	**115**	99	**86%**
25	44.3	65	872	115							
26	47.7	70	291	38							
27	49.4	72.5	0	0							

The gross power is equal to the thrust times the velocity at which it is delivered, or the prop speed. Clearly, this engine can't produce 8433 pounds of thrust and 1112 hp. Recall that Equation 6.3 assumed there was no prop slip. If we include enough slip so that the gross power delivered is equal to 115 hp, we get columns E through K.

The net power is equal to the thrust times the forward velocity, or the boat speed. The boat on which this example is based tops out at 42.5 mph. If the engine develops 115 hp, then the thrust would be 872 pounds and the net would be 99 hp. At top speed the prop slip would only be 3% and the prop efficiency would be 86%, as shown in the table. This next figure shows the prop speed and slip over the range of boat speed (at full-throttle, of course):

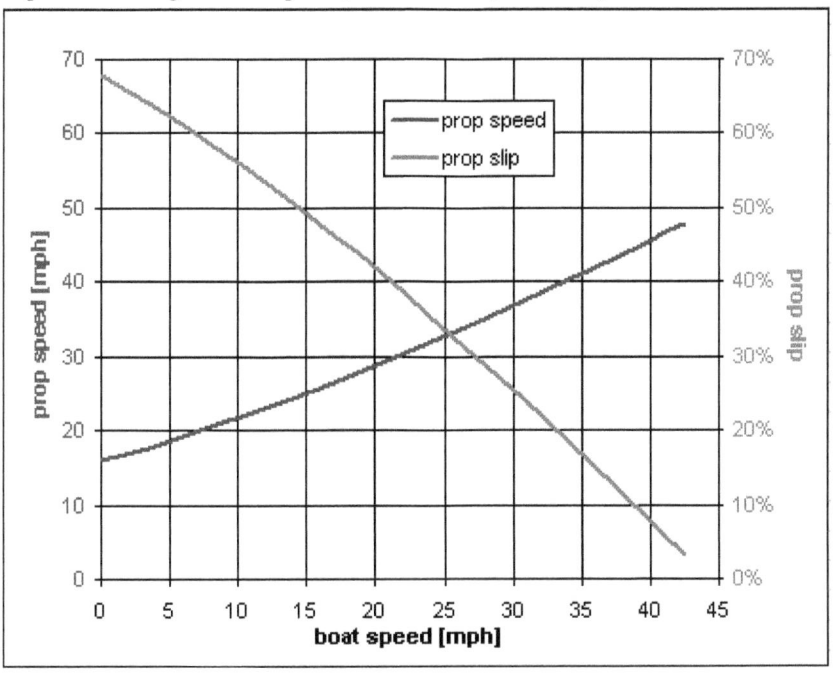

The thrust is proportional to the area times the velocity times the change in velocity and the power is equal to the thrust times the velocity, so the power required to turn a propeller is proportional to $D^5\Omega^3$. The power is also proportional to the density, which is airplane propellers are so much larger than marine props.

$$P \propto \rho D^5 \Omega^3 \qquad (6.5)$$

If you want to go fast, you must throw water even faster. As the engine can only deliver so much power, you must throw a lot less water faster, which means a much smaller, faster rotating prop and a 1:1 gear ratio. This next figure illustrates this relationship of scale:

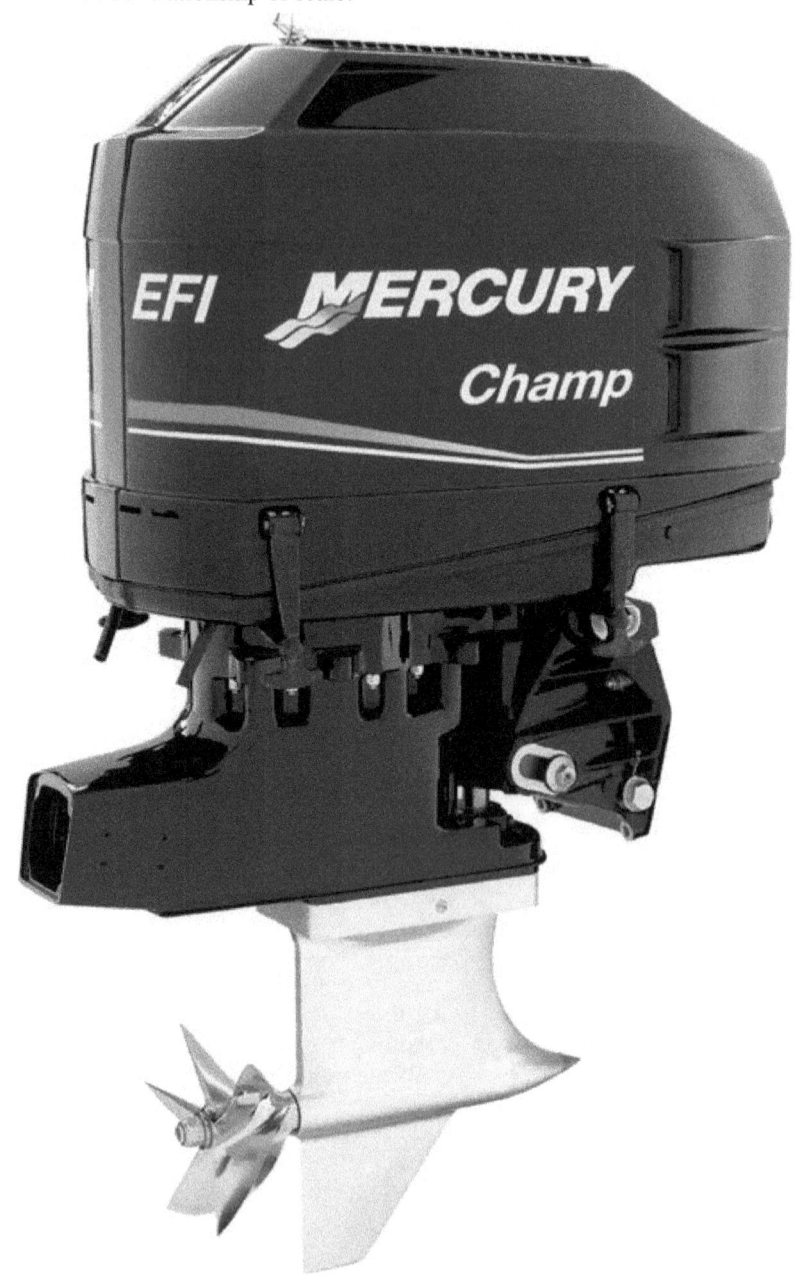

Various modifications have been developed to enhance certain aspects of marine propellers. Avoidance and control of cavitation has been the motivation for much of this effort. Cavitation not only robs power, it causes considerable damage, as illustrated in this next figure:

Photo of a blade damaged by Cavitation Erosion

The most common modification to control cavitation is called *cupping*. Traditional cupping involved slightly rolling up the trailing edge, as illustrated in the following figure:

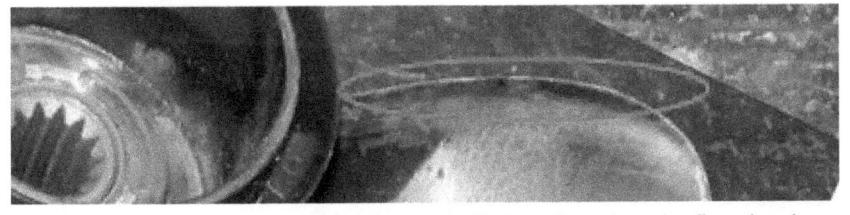

The purpose of this rolled edge is similar to the wing tip fins that have become quite common on commercial aircraft. These were originally added to large planes to discourage vortices from rolling off the end of the wing and possibly upsetting the flight of smaller aircraft. Once the airlines discovered that these wingtips slightly improved performance, they started putting them on planes of all sizes. Boat racers knew this all along.

"If a little is good, then a lot must be better," has always been a tendency with me. I applied this reasoning to cupping props. It has become quite popular, but I started the practice in 1966. This next figure shows the "new" way of cupping a prop:

The rolled edge is much deeper and extends all the way around the end of the blade. I'm sure there are more elegant ways of accomplishing this, but I used a trailer hitch ball and a hammer. I shared this secret with a friend, Bruce Borkenhagen. He won an outboard regatta championship and the other racers had to know why. He kept his prop covered, which only increased the other's curiosity. Bruce went on to work for OMC racing and testing engines. I'm glad the secret got out. Someone has applied my reasoning and taken things a bit further.

Consider the following diagram showing the ideal slipstream around a propeller:

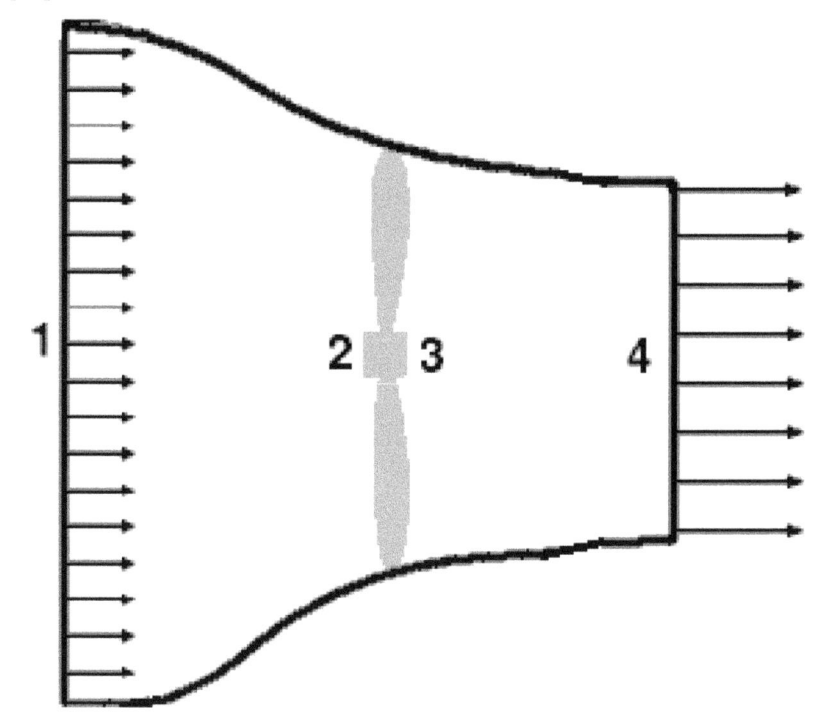

The conservation of mass (or volume) gives us:

$$V_2 = V_3 = V_P = \frac{V_1 + V_4}{2} \tag{6.6}$$

Efficiency is the ratio of the speed at which the force is delivered divided by the speed at which it is applied:

$$\eta = \frac{V_1}{V_P} \tag{6.7}$$

The power is equal to the thrust times the delivery speed:

$$P = \frac{\pi D^2 \rho}{4} V_1 \left(V_4^2 - V_1^2 \right) \tag{6.8}$$

Taking the partial derivative of Equation 6.8 with respect to V_1 and setting this to zero gives us the point of optimum performance:

$$\frac{\partial P}{\partial V_1} = \frac{\pi D^2 \rho}{4}\left(V_4^2 - 3V_1^2\right) = 0 \qquad (6.9)$$

The optimum performance occurs at:

$$\frac{V_4}{V_1} = \sqrt{3} \qquad (6.10)$$

Combining Equations 6.6, 6.7, and 6.10 yields:

$$\eta_{opt} = \frac{2}{1+\sqrt{3}} = 73.2\% \qquad (6.11)$$

The maximum efficiency may be 1, occurring when $V_1=V_P$, but there's no thrust at this point, so this case is of no interest. Consider this: If you want to go 60 mph, you need to plan on pushing water out the back at 104 mph, which will require considerable rpm and pitch.

Chapter 7. Axial Fans

Axial fans are similar to propellers in many ways. We will consider only large shrouded fans moving air in this chapter, as we considered only unshrouded marine propellers in the last. The following figure shows a typical fan atop a cooling tower along with a man standing on the hub to convey the scale:

Such fans are often 30 feet in diameter and are driven by electric motors developing around 200 hp or more.

The following is a typical set of efficiency curves for such a fan:

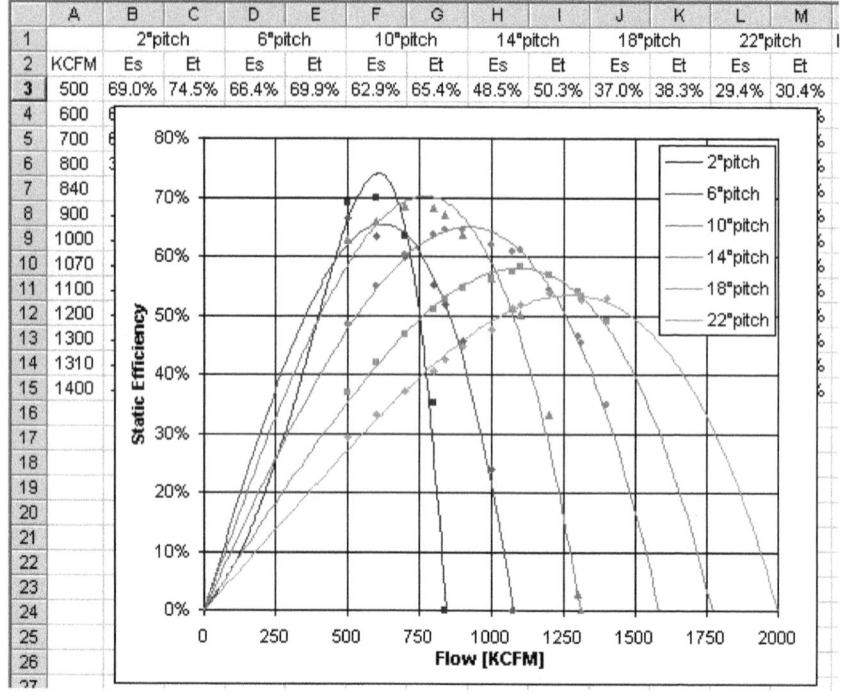

	A	B	C	D	E	F	G	H	I	J	K	L	M
1		2"pitch		6"pitch		10"pitch		14"pitch		18"pitch		22"pitch	
2	KCFM	Es	Et	Es	Et	Es	Et	Es	Et	Es	Et	Es	Et
3	500	69.0%	74.5%	66.4%	69.9%	62.9%	65.4%	48.5%	50.3%	37.0%	38.3%	29.4%	30.4%
4	600												
5	700												
6	800												
7	840												
8	900												
9	1000												
10	1070												
11	1100												
12	1200												
13	1300												
14	1310												
15	1400												

The curves are the manufacturer's advertised performance and the points are test data (i.e., measured results). For this type of fan the pitch angle is most often reported. This angle is in reference to an indicator bolted to each blade and is only approximate. The actual angle changes over the length of the blades. In the literature for this type of fan it is common to refer to both *static* and a *total* efficiency. Static efficiency is the only one that has meaning. Total efficiency is calculated by adding the velocity head ($\rho V^2/2$) to the pressure rise developed across the fan. It's just a bigger number that sounds better. If we divide the X-axis by the stall speed times the area and the Y-axis by the maximum value, the data collapse to a single curve:

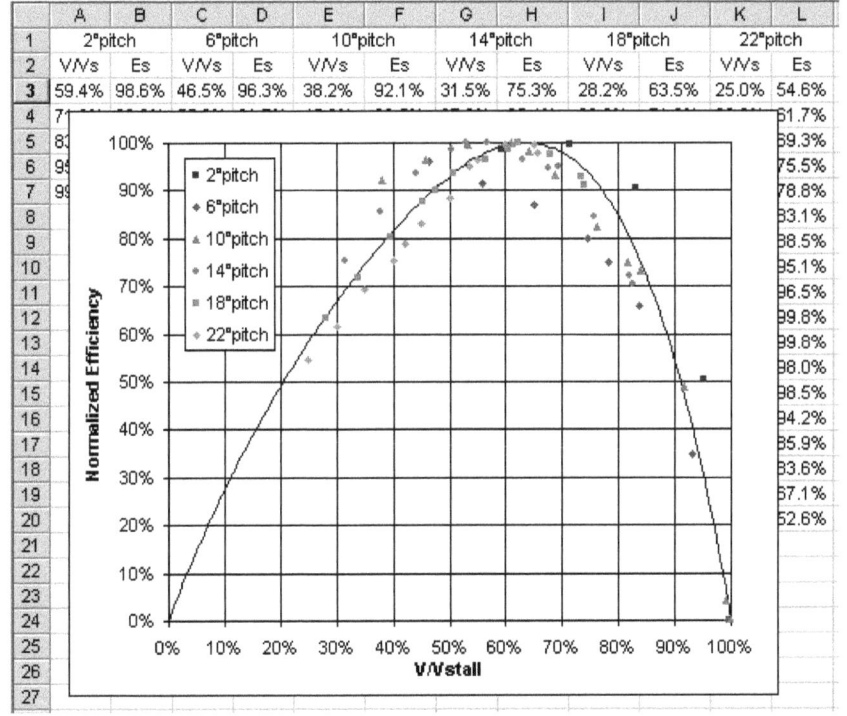

	A	B	C	D	E	F	G	H	I	J	K	L
1	2"pitch		6"pitch		10"pitch		14"pitch		18"pitch		22"pitch	
2	V/Vs	Es	V/Vs	Es	V/Vs	Es	V/Vs	Es	V/Vs	Es	V/Vs	Es
3	59.4%	98.6%	46.5%	96.3%	38.2%	92.1%	31.5%	75.3%	28.2%	63.5%	25.0%	54.6%

(chart: Normalized Efficiency vs V/Vstall for 2", 6", 10", 14", 18", 22" pitch)

The stall speed is the point where there is no pressure rise across the fan. This is equivalent to a propeller advancing at $\Phi\Omega$ and producing no thrust. Of course, you can't achieve that condition with a boat unless there's another one pulling it and you can't achieve this condition with a fan unless you have second fan pulling the air through the first one. This is the point where the *total* efficiency of the propeller and fan are both equal to 100% (the *static* efficiency is zero); yet produce no thrust.

The above curve for normalized efficiency vs. $\beta = V/V_{stall}$ is given by:

$$\frac{\eta}{\eta_{max}} = \frac{3}{4}\beta(1-\beta)((9\beta-3)\beta+4) \qquad (7.1)$$

The maximum efficiency for an axial fan is limited by the swirl, as this consumes power, but doesn't produce thrust. This can be approximated by the following integral:

$$\eta_{swirl} = \int_0^1 \cos(\arctan(\varphi/x))\,dx \qquad (7.2)$$

In the preceding equation $\varphi = \Omega/\pi D$ is the pitch ratio. This integral can be computed numerically and approximated by the following expression:

39

$$\eta_{swirl} = e^{\left[\frac{-1.17583\varphi}{1+(0.324221-0.00320923\varphi)\varphi}\right]} \quad (7.3)$$

The maximum pressure ratio, $\gamma_{max}=\Delta P_{max}/(\rho V^2/2)$, for an axial fan can be estimated by the following equation, which is based on a variety of test data for this type of fan.

$$\gamma_{max} = \frac{0.23965}{\varphi^{1.57314}} \quad (7.4)$$

The normalized pressure ratio can be computed with this next equation:

$$\frac{P}{P_{max}} = 1 - \varphi^3 \quad (7.5)$$

The static pressure difference developed by the fan pictured and the motor power required to turn it are shown in this next figure:

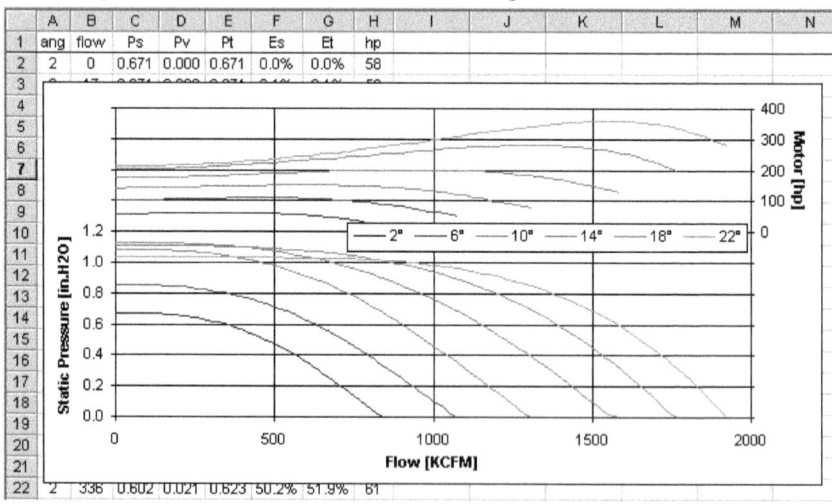

These curves were calculated using Equations 7.1 through 7.5 and are in excellent agreement with the manufacturer's published performance curves. The static pressure curves are at the bottom and are read off the left scale. The power curves are at the top and are read off the right scale. This curve format is typical for the industry.

Just as propeller designs vary considerably, so do axial fan designs. Howden makes the most interesting blade shapes. This figure is from their on-line brochure.

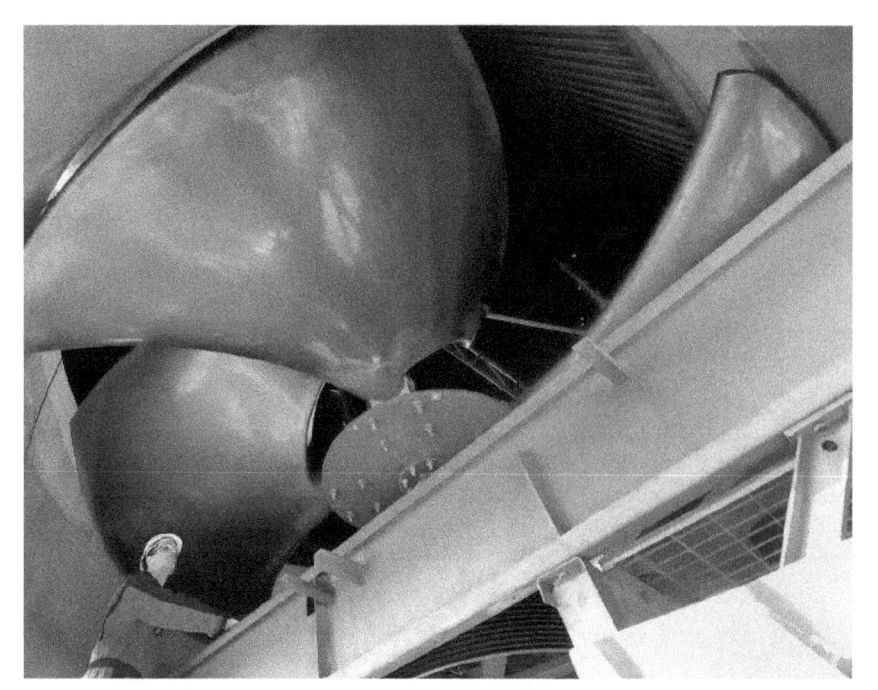

These fans are noticeably quieter than the common variety shown in the previous figure. They may have some performance advantage at certain operating conditions as well.

Chapter 8. Rogue Waves

I survived a rogue wave. The year was 1967 and I was 15 years old. My family was on a 38-foot ketch (two-masted sailboat) with friends. It was a beautiful day and we were just 5 miles off the coast of Ft. Lauderdale. A squall sprang up, swept over us, and quickly passed. It seemed a trifle and we would have forgotten it entirely, but then there was the wave.

We all saw it, but no one said a word. We just stared and clung to the boat. At first I thought it an optical illusion, for the crest of the wave was above the mast. As it approached, the boat rose and the rapid lift added a strong downward pull that made us feel even smaller in the presence of this towering swell. We thought the upward acceleration was rapid until the wave passed under us and we dropped into the trough.

When the boat landed, the bottom split open and we could see light along the keel. We hastily stuffed clothes into the open seams, but the bilge filled quickly and the engine died in less than a minute. We had just enough time to call for help on the shortwave before the batteries and electrical system gave in to the salt water.

The ketch was unusually sturdy, made entirely of teak. It had been specially designed and built by two men who intended to sail around the world in it. Before it was finished, they had a falling-out and sold the boat to my parents' friends, who named it the Mona VI. If the boat had been larger, the wave would have swept over us. Instead, we floated over it like a cork. If the boat had not been as sturdy, we would have sank immediately.

The Coast Guard station at nearby Port Everglades dispatched a helicopter, which arrived on the scene in less than 20 minutes. They dropped a 55-gallon drum containing a large pump, which we rapidly deployed. It was a close race against the sea, but the pump finally prevailed and we didn't sink.

A nearby 65-foot sport fisherman, named Kitty Kat, heard our call to the Coast Guard and also came to our rescue. They slowly towed us into port and to a marina, where the Mona VI was hauled up and the damage assessed. The hull was split in several places, not just along the keel and several of the ribs were cracked, but she could be repaired. The hardest part would be acquiring the teak.

We are not the only ones to have such an experience. Many have not lived to tell the story. As far as we know, no one else saw the wave. We don't know what it did when it hit the beach, but there were no reports of drowning that day. There were many other boats in this busy coastal waterway, but only we rode the rogue. For more stories and further reading, I highly recommend Susan Casey's *The Wave*. The following is a quote from the Wikipedia rogue wave page:

> *One of the remarkable features of the rogue waves is that they always appear from nowhere and quickly disappear without a trace.*

The history of the sea is filled with stories of rogue waves, told by sailors who survived to tell the tale. Rogue waves have even struck objects on land, such as a lighthouse. Still, there were those entirely skeptical of their existence. It was not until recent years that sufficient photographic and instrument evidence has become available to finally convince the skeptics.

Rogue wave [image from BBC Horizons, 2002]

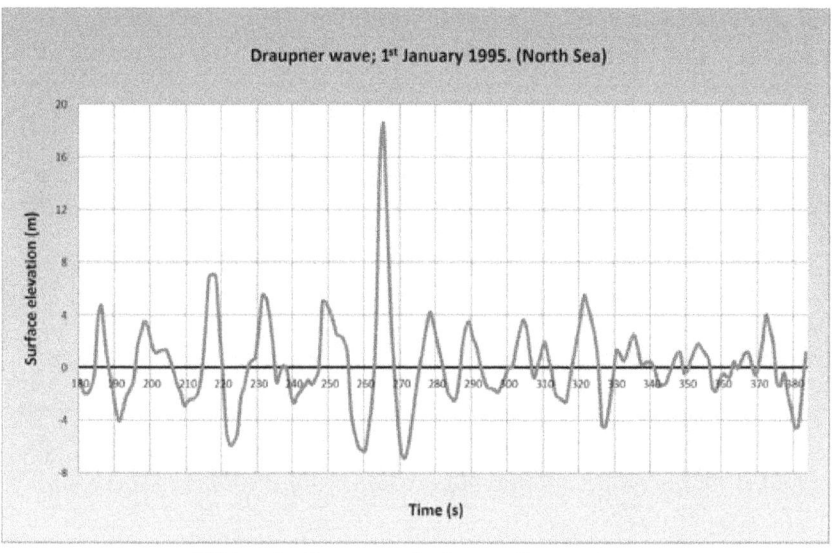

Wikimedia Commons, the free media repository

The first rogue wave to be detected by a measuring instrument occurred at the Draupner platform in the North Sea off the coast of Norway on January 1, 1995. The surrounding waves had a height of approximately 12 meters (39 ft);

whereas, the rogue had a height of 25.6 meters (84 ft). Prior to that measurement, no instruments had recorded a rogue wave.

There are several mathematical approximations for rogue waves. The most common is known as the solitary wave, which is defined thus:

$$\eta(x,t) = h \sec h^2\left(\frac{x-ct}{\lambda}\right)$$

$$\lambda = \sqrt{\frac{4d^3}{3h}} \quad (8.1,2,3)$$

$$c = \sqrt{g(d+h)}$$

In these equations η is the surface elevation, x is the distance from the crest, t is time, h is the height, *sech* is the hyperbolic secant (equal to 1/*cosh*, the hyperbolic cosine), λ is the wavelength, d is the depth, c is the celerity (i.e., wave speed), and g is the acceleration of gravity. This is what the wave profile looks like in one dimension:

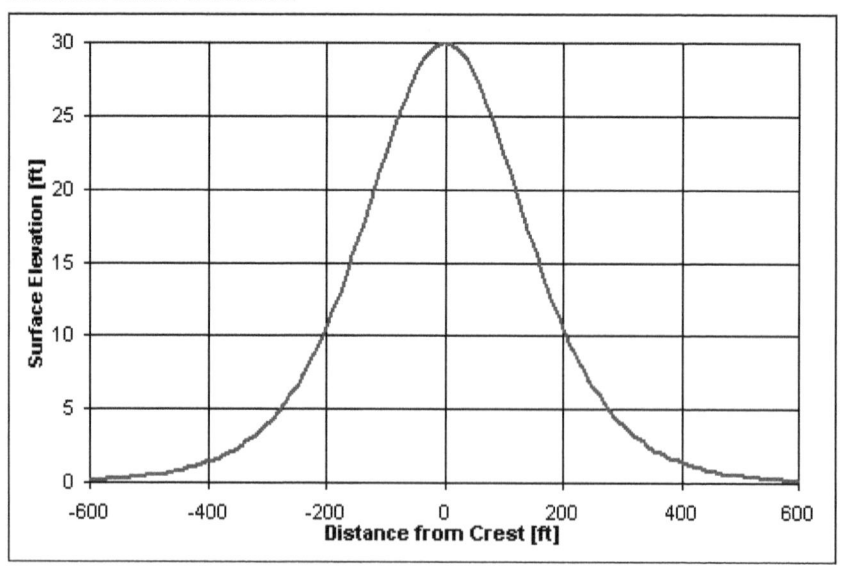

This is what the wave profile looks like in two dimensions (distance and time):

Waves can travel very long distances at sea and may crash on a distant shore. They can reflect off objects, but eventually dissipate. As they are created in many different locations and travel in different directions, they may occasionally combine in spot for a brief time and form a rogue wave. The following figure is exaggerated, but does provide a rough picture of how multiple waves coming from different directions might combine to create a rogue.

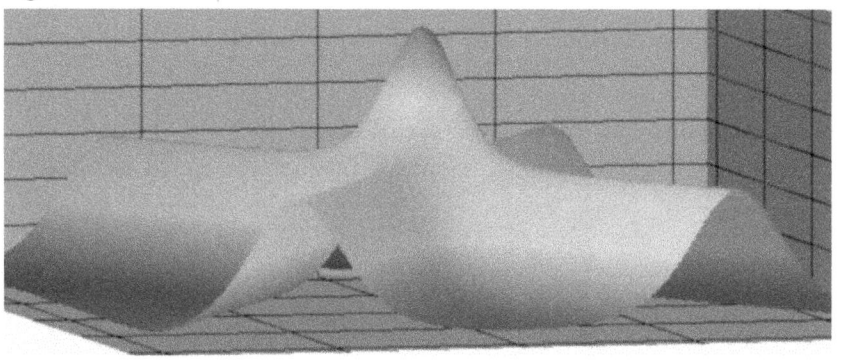

Chapter 9. The Heat Index

The heat index (or "humiture") was developed by George Winterling in 1978.[3] It was adopted by the National Weather Service a year later. It is derived from work carried out by Robert G. Steadman.[4,5] The heat index contains assumptions about the human body mass and height, clothing, amount of physical activity, thickness of blood, sunlight and ultraviolet radiation exposure, and the wind speed. In other words, it is based on empirical data and attempts to capture the perception of temperature as it pertains to humans and their environment. The following table is provided by the National Oceanographic and Atmospheric Administration (NOAA):

Tmp (°F)	\multicolumn{21}{c}{The Heat Index — Relative Humidity}
	0%	5%	10%	15%	20%	25%	30%	35%	40%	45%	50%	55%	60%	65%	70%	75%	80%	85%	90%	95%	100%
135	120	126																			
130	117	122	131																		
125	111	116	123	131	141																
120	107	111	116	123	130	139	148														
115	105	107	111	115	120	127	135	143	151												
110	99	102	105	108	112	117	123	130	137	143	150										
105	95	97	100	102	105	109	113	118	123	129	135	142	149								
100	91	93	95	97	99	101	104	107	110	115	120	126	132	138	144	150					
95	87	88	90	91	93	94	96	98	101	104	107	110	114	119	124	130	136	140	150		
90	83	84	85	86	87	88	90	91	93	95	96	98	100	102	106	109	113	117	122	126	131
85	78	79	80	81	82	83	84	85	86	87	88	89	90	91	93	95	97	99	102	105	108
80	73	74	75	76	77	77	78	79	79	80	81	81	82	83	84	85	86	87	88	89	90
75	69	69	70	71	72	72	73	73	74	74	75	75	76	76	77	77	78	78	79	79	80
70	64	64	65	65	66	66	67	67	68	68	69	69	70	70	70	70	71	71	71	71	72

■ = Heatstroke risk extremely high
■ = Heat exhaustion likely, heatstroke possible
■ = Heat exhaustion possible
■ = Fatigue possible

There are several articles on the Web that provide formulas for calculating the heat index. These typically include both a short and long version, as listed in the table on the next page. These work well enough, but are overly complicated, as they are the result of a multivariate regression on temperature and relative humidity. Relative humidity is a convenient term, but not a particularly useful one when it comes to thermodynamic properties.

It is far more revealing to consider moisture content. While relative humidity varies from 0 to 100%, moisture content varies by orders of magnitude. Warm air doesn't just hold a little more moisture; it can hold a whole

---

[3] There are many articles on the Web that say Winterling's work was released in 1978, but no proper citation. These all refer back to the same article, which provides a much more recent citation: "A Lifelong Passion For Weather WJXT," George Winterling, April 23, 2009.

[4] "The Assessment of Sultriness. Part I: A Temperature-Humidity Index Based on Human Physiology and Clothing Science," R. G. Steadman, Journal of Applied Meteorology, July 1979, Vol 18 No7, pp. 861–873.

[5] "The Assessment of Sultriness. Part II: Effects of Wind, Extra Radiation and Barometric Pressure on Apparent Temperature," R. G. Steadman, Journal of Applied Meteorology, July 1979, Vol 18 No7, pp. 874–885.

lot more moisture, as illustrated in the following figure of moisture content at saturation (i.e., 100% relative humidity) vs. temperature.

The dry-bulb temperature is what you would measure with an ordinary thermometer. There is also a wet-bulb temperature that is measured by putting a little moist sock on the bulb of the thermometer and gently blowing air over it. The latter is more like what you feel with your skin, which is also moist and porous.

Short Version

$HI=-42.38+2.049*T+10.14*RH$
$-0.2248*T*RH-0.006838*T^2$
$-.05482RH^2+0.001229*T^2*RH$
$+0.0008528*T*RH^2-0.00000199$
$*T^2*RH^2$

Long Version

$HI=16.923+((1.85212E-1)*T)$
$+(5.37941*RH)$
$-((1.00254E-1)*T*RH)$
$+((9.41695E-3)*T^2)$
$+((7.28898E-3)*RH^2)$
$+((3.45372E-4)*T^2*RH)$
$-((8.14971E-4)*T*RH^2)$
$+((1.02102E-5)*T^2*RH^2)$
$-((3.8646E-5)*T^3)$
$+((2.91583E-5)*RH^3)$
$+((1.42721E-6)*T^3*RH)$
$+((1.97483E-7)*T*RH^3)$
$-((2.18429E-8)*T^3*RH^2)$
$+((8.43296E-10)*T^2*RH^3)$
$-((4.81975E-11)*T^3*RH^3)$

As shown in the table below, at 188.4°F there is as much water vapor as air (i.e., water vapor/air mass ratio = 1).

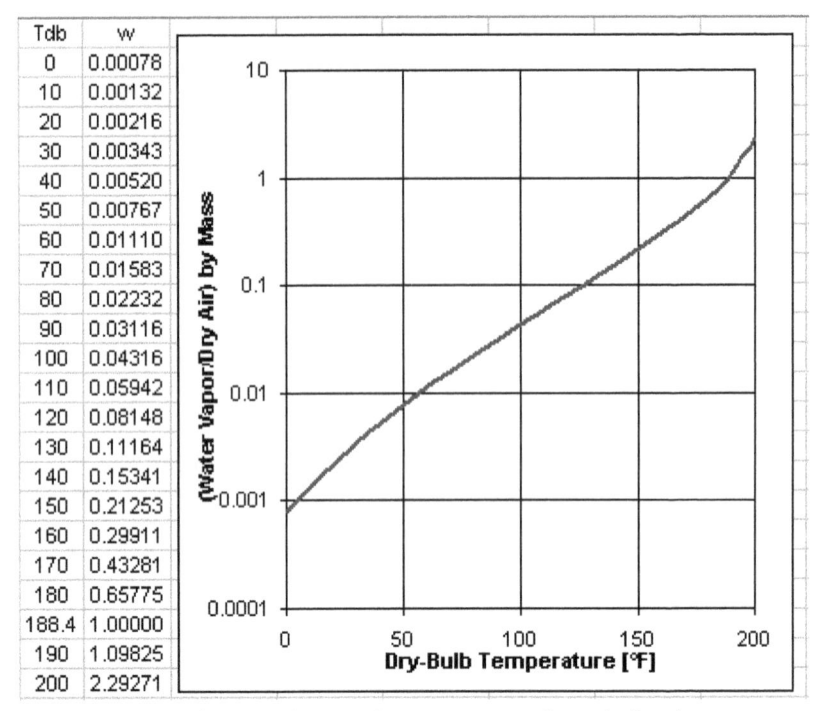

Tdb	w
0	0.00078
10	0.00132
20	0.00216
30	0.00343
40	0.00520
50	0.00767
60	0.01110
70	0.01583
80	0.02232
90	0.03116
100	0.04316
110	0.05942
120	0.08148
130	0.11164
140	0.15341
150	0.21253
160	0.29911
170	0.43281
180	0.65775
188.4	1.00000
190	1.09825
200	2.29271

Water vapor also contains much more energy than air for the same mass. This next table shows the fraction of the total energy contained in the water vapor alone.

Fraction of Total Energy in Water Vapor Alone

Tdb °F	0%	10%	20%	30%	40%	50%	60%	70%	80%	90%	100%
30	0%	5%	9%	13%	17%	20%	23%	26%	29%	31%	34%
40	0%	5%	10%	15%	19%	23%	26%	29%	32%	34%	37%
50	0%	6%	12%	17%	22%	26%	29%	33%	36%	38%	41%
60	0%	8%	14%	20%	25%	29%	33%	37%	40%	43%	46%
70	0%	9%	17%	23%	29%	34%	38%	42%	45%	48%	51%
80	0%	11%	20%	27%	33%	38%	43%	47%	50%	53%	56%
90	0%	13%	23%	31%	38%	44%	48%	52%	56%	59%	61%
100	0%	16%	27%	36%	43%	49%	54%	58%	61%	64%	66%

The blue numbers above the stair-step line are where there is more energy in the air than the water vapor and the red ones below are where there is more energy in the water vapor than air. Rather than performing a regression for heat index on temperature and relative humidity, this should be done on temperature and absolute humidity. The latter is the ratio of water vapor to air by mass and given the symbol *w*. This regression is much simpler than the preceding two.

```
HI=-16.2746+(1.44629-0.00340438*Tdb)*Tdb
   +(-2478.16+25.2577*Tdb+52181.1*w)*w
```

The agreement between this regression and 996 tabulated values is shown in the following figure:

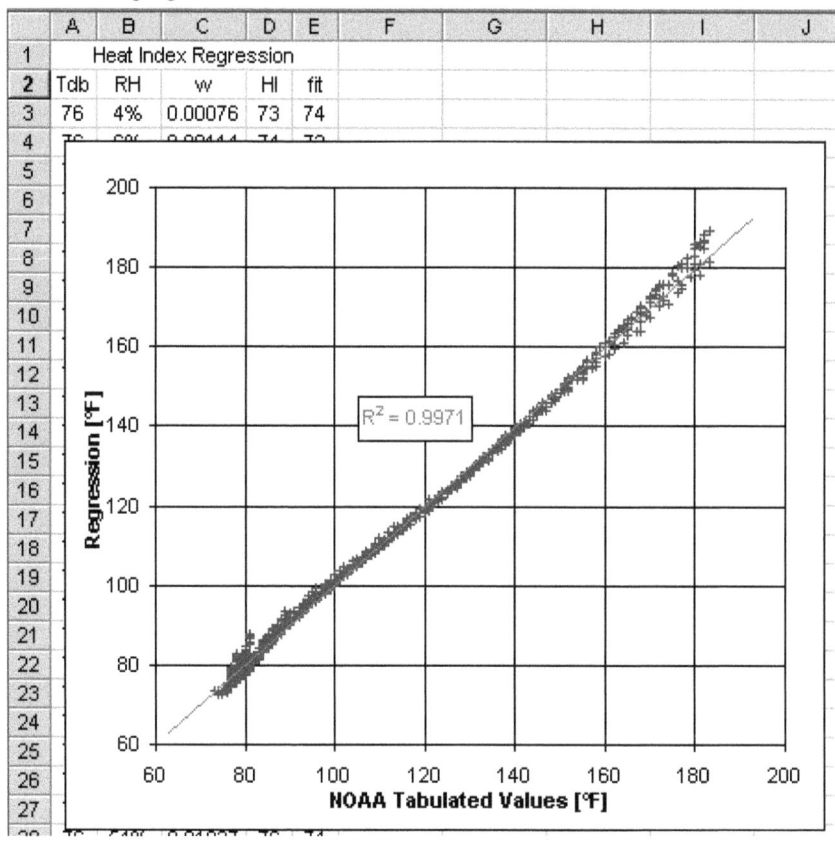

	A	B	C	D	E	F	G	H	I	J
1	Heat Index Regression									
2	Tdb	RH	w	HI	fit					
3	76	4%	0.00076	73	74					

Regression [°F] vs NOAA Tabulated Values [°F], $R^2 = 0.9971$

49

Chapter 10. Wind Chill

Wind chill is the apparent reduction in temperature accompanying the flow of air–primarily as it pertains to humans exposed to cold weather. This effect is due to evaporation and so it only applies to porous surfaces that are moist (i.e., skin). There are many formulas and tables for wind chill. The first of these was developed by Paul Allman Siple and Charles F. Passel, based on their experience during Antarctic exploration in the 1930s. Their work was updated and adopted by the National Weather Service by the 1970s, which is the source of the chart below:

mph	40	35	30	25	20	15	10	5	0	-5	-10	-15	-20	-25	-30	-35	-40	-45
5	36	31	25	19	13	7	1	-5	-11	-16	-22	-28	-34	-40	-46	-52	-57	-63
10	34	27	21	15	9	3	-4	-10	-16	-22	-28	-35	-41	-47	-53	-59	-66	-72
15	32	25	19	13	6	0	-7	-13	-19	-26	-32	-39	-45	-51	-58	-64	-71	-77
20	30	24	17	11	4	-2	-9	-15	-22	-29	-35	-42	-48	-55	-61	-68	-74	-81
25	29	23	16	9	3	-4	-11	-17	-24	-31	-37	-44	-51	-58	-64	-71	-78	-84
30	28	22	15	8	1	-5	-12	-19	-26	-33	-39	-46	-53	-60	-67	-73	-80	-87
35	28	21	14	7	0	-7	-14	-21	-27	-34	-41	-48	-55	-62	-69	-76	-82	-89
40	27	20	13	6	-1	-8	-15	-22	-29	-36	-43	-50	-57	-64	-71	-78	-84	-91
45	26	19	12	5	-2	-9	-16	-23	-30	-37	-44	-51	-58	-65	-72	-79	-86	-93
50	26	19	12	4	-3	-10	-17	-24	-31	-38	-45	-52	-60	-67	-74	-81	-88	-95
55	25	18	11	4	-3	-11	-18	-25	-32	-39	-46	-54	-61	-68	-75	-82	-89	-97
60	25	17	10	3	-4	-11	-19	-26	-33	-40	-48	-55	-62	-69	-76	-84	-91	-98
Frostbite Times			30 minutes					10 minutes				5 minutes						

NOAA Wind Chill Chart - Temperature [°F]

This chart is based on the following empirical relationship, where T_{WC} is the wind chill, T_{DB} is the dry-bulb temperature, and V is the wind speed in mph.

$$T_{WC} = 35.74 + 0.6215 T_{DB} \\ + (0.4275 T_{DB} - 35.75) V^{0.16} \tag{10.1}$$

This formula has the wind chill increasing with the 4/25ths power of wind speed, which leads to the following question. If you're wearing a wet T-shirt on a hot day, you feel cooler until the shirt dries out, because the water is evaporating. If there's wind, it feels cooler than with no wind. The SR-71 Blackbird gets so hot at Mach 3.2 that the designers had to allow for thermal expansion. The accommodation for thermal expansion in this aircraft is enough that the fuel system leaks when it's sitting on the runway. It's only natural to wonder if this effect just keeps getting more so–regardless of how fast the wind is blowing–or if there comes a point when the wind blowing on your wet T-shirt would make you hotter.

You can readily find where people have asked this question on the Web, but try finding a reasoned and polite answer. Mr. Smarty-Pants replies, "Of course it get hotter as you go faster. You would burn up from friction heat at 15 km/s

[33,500 mph]." At that speed evaporative cooling vs. friction heating would be the least of our problems. Mr. Know-It-All writes, "Spacecraft heat up because they compress air in front of them, and not because of friction as they pass through the air." This is technically correct, but not helpful.

Stagnation heating is due to kinetic energy being converted into thermal energy as the wind impacts an object. The increase in temperature due to this phenomenon is given by:

$$\Delta T = \frac{V^2}{2C_P} \qquad (10.2)$$

Equations 10.1 and 10.2 can combined, used to extend the preceding table, and produce the following graph:

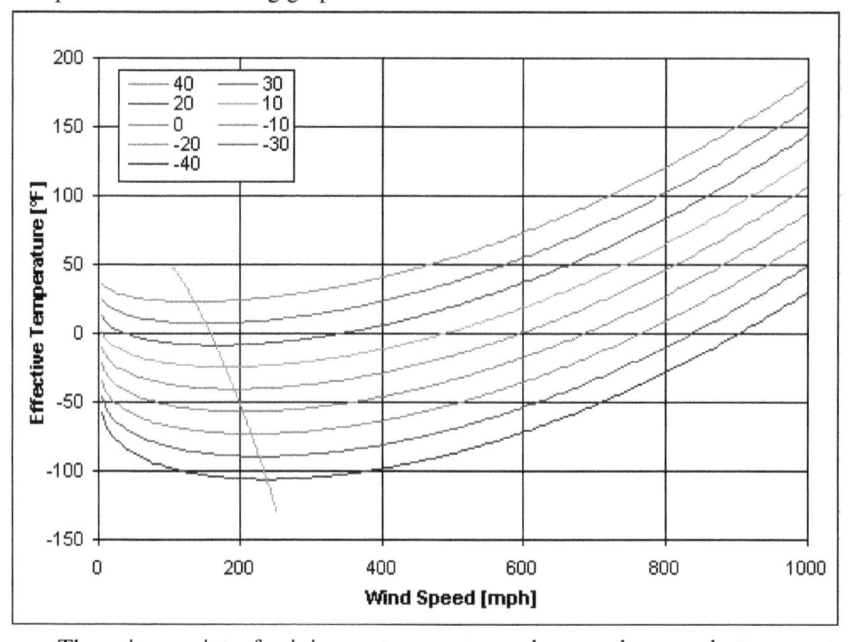

There is a point of minimum temperature where each curve bottoms-out. The magenta line is drawn through the minimum points, which range from 135 to 243 mph as the temperature goes from 40°F to -40°F. This isn't a precise calculation, but it does show what most of us have wondered: there is an optimal or minimum point.

Chapter 11. Heating, Cooling, and Entropy

Consider an 8-ounce cup of coffee. Let's say it starts out at 195°F, reportedly the perfect brewing temperature. The room is held constant at 75°F. Rather than drinking it right away, you let it cool. After 30 minutes the temperature of your coffee has dropped to 90°F. This is a classic heat transfer problem, often given for homework or on a test. The solution is straightforward:

$$T = 75 + 120\, e^{\frac{-t}{\tau}}$$
$$\tau = \frac{30}{\ln(8)} \quad (10.1)$$

Eventually, the coffee will reach the temperature of the room. What is the change in entropy for this process? In order to simplify things, we will ignore the cup and assume the coffee is just water. The change in entropy for the coffee is given by:

$$\Delta S_{coffee} = mC_P \ln\left(\frac{T_\infty}{T_0}\right) \quad (10.2)$$

In this case m is the mass (8 oz.), C_P is the specific heat (1 BTU/lbm/°F), T_0 is the initial temperature (195°F or 655°R), and T_∞ is the final temperature (75°F or 535°R). The change in entropy for the coffee, Δs, is equal to -0.101184 BTU/°R. This is a decrease in entropy (i.e., a negative change), but this is only for the coffee. We must also consider the change in entropy for the room. Here we apply the Second Law of Thermodynamics for a closed system.[6]

$$\Delta S_{room} = \frac{\Delta Q}{T_{room}} = \frac{mC_P(T_0 - T_\infty)}{T_\infty} \quad (10.3)$$

The mass, specific heat, and temperatures are the same as before and the result is +0.112150 BTU/°R. The combined result for the coffee and room is +0.010965 BTU/°R so that there is a net increase in entropy for this cooling process.

Next consider an 8 oz. cold beverage that begins at 35°F (495°R) and is allowed to warm up to room temperature. The same formulas apply; only the initial temperature is different. The change of entropy for the cold beverage is

---

[6] A *closed* system is one in which there is no exchange of mass with its surroundings (i.e., the coffee stays in the cup). An *isolated* system is one in which there is no exchange of mass or energy (e.g., an insulated tank) with the surroundings. An *open* system is one in which there is an exchange of mass and energy with the surroundings. People often refer to something as a *closed* system when they should be calling it an *isolated* system. The two are not interchangeable.

+0.0388545 BTU/°R and for the room is -0.0373832 BTU/°R, making the net for this warming process +0.001471 BTU/°R. Whether heating or cooling, there's always an increase in entropy. We can calculate this for a range of initial temperatures in a simple spreadsheet.

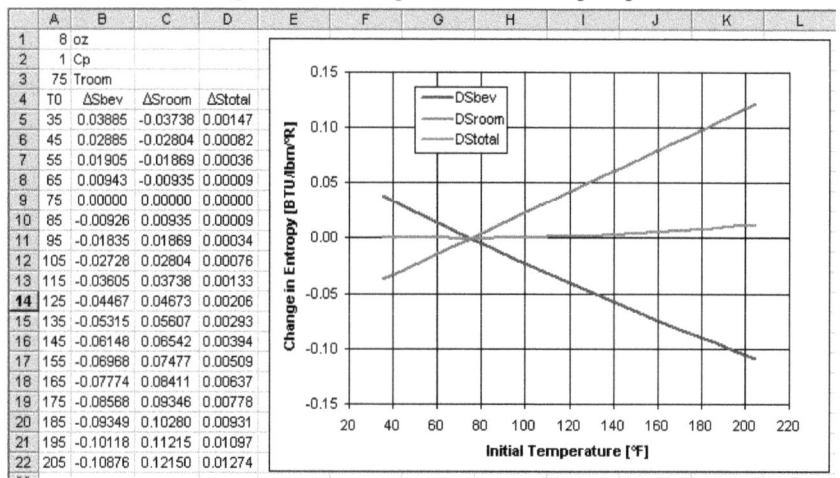

	A	B	C	D
1	8	oz		
2	1	Cp		
3	75	Troom		
4	T0	ΔSbev	ΔSroom	ΔStotal
5	35	0.03885	-0.03738	0.00147
6	45	0.02885	-0.02804	0.00082
7	55	0.01905	-0.01869	0.00036
8	65	0.00943	-0.00935	0.00009
9	75	0.00000	0.00000	0.00000
10	85	-0.00926	0.00935	0.00009
11	95	-0.01835	0.01869	0.00034
12	105	-0.02728	0.02804	0.00076
13	115	-0.03605	0.03738	0.00133
14	125	-0.04467	0.04673	0.00206
15	135	-0.05315	0.05607	0.00293
16	145	-0.06148	0.06542	0.00394
17	155	-0.06968	0.07477	0.00509
18	165	-0.07774	0.08411	0.00637
19	175	-0.08568	0.09346	0.00778
20	185	-0.09349	0.10280	0.00931
21	195	-0.10118	0.11215	0.01097
22	205	-0.10876	0.12150	0.01274

More than once I have been asked, "Instead of wasting all this energy, why don't we just build a car or power plant that is 100% efficient?" That would be nice, wouldn't it? Too bad it's not possible. I even went through this derivation for one such individual. It was a waste of time.

At least he did eventually suggest that we simply operate at the minimum point where the red curve goes through zero. That's also the point where nothing happens. Your beverage is neither hot nor cold. A car up on blocks in the front yard doesn't use any gas, but it doesn't provide transportation either. If you want to make something happen, you're going to generate entropy. If you want it to happen fast, you're going to generate a lot of entropy. You can't undo the process without generating more entropy.

This brings us to the topic of irreversibility. In this example, the irreversibility is equal to the net change in entropy times the temperature of the room. A graph of irreversibility would look the same, only multiplied by 535°R.

Irreversibility is easier to understand than a change in entropy because it is equal to the *lost work*.[7] If we had a perfectly efficient little heat engine, we could have extracted 5.86643 BTU of work from our cup of coffee as it cooled and 0.787153 BTU of work from our cold beverage as it warmed. The lost work for

---

[7] Heat is the transient form of energy that crosses a system boundary by virtue of a temperature difference. No ΔT = no heat. Work is the transient form of energy that crosses a system boundary by virtue of a force. No force = no work.

the first process is 7.45 times as much as for the second, which is why we make power from burning stuff rather than towing icebergs down south to melt them.

Chapter 12. The Perfect Power Plant

Have you ever noticed in cop shows how the star comes up behind a technician (expendable person in a single episode) who's staring at a traffic cam video looking for the bad guy? The star says, "There... that's a license plate. Zoom in on that." Just once I'd like the technician to say, "I've been staring at this video all day and didn't notice that license plate. I never would have thought to zoom in on it if you hadn't suggested it. I've only been working at this for twenty years. You'd think by now I'd get the hang of this stuff."

What does this have to do with the ideal power plant? More than once I've been asked, "Why not just build a perfectly efficient power plant? That way there will be no pollution." In spite of spending years in school taking courses others avoid and drawing upon decades of experience, engineers are really stupid. Like the video technician, they're waiting for a star that can't brush their own hair to come along and tell them how to do their job better. Before we analyze this suggestion, let's consider a few facts.

Bull Run Steam Plant was built near the end of an era of coal plant expansion, coming online in 1967. It was high tech at the time, the largest plant in the world. It won many awards and was the most efficient power plant in North America for over a decade. It still produces enough power for 600,000 homes and businesses.

In 1977 Bull Run was retrofitted with electrostatic precipitators to remove fly ash from the smoke stack. This cleaned up the exhaust, but the retrofit cost more than the whole plant did to build in the first place. What if you found out a new muffler cost more than the car and that you had to replace it? In 2008 Bull Run was fitted with a flue gas desulphurization (FGD) system (i.e., a wet scrubber) that cost another $450M.

The star of our show may want to know why there's such a problem with the exhaust in the first place. It's because Bull Run burns 7,300 tons of coal a day to power those 600,000 homes. That's a trainload a day. Yes, a train comes and dumps a load there almost every day of the year and sometimes twice a day. It's a lot cleaner and more efficient than if those 600,000 homes burned that coal in their fireplace, not to mention how many trees it would take to burn wood instead of coal.

The star of our show may suggest, "Let's quit burning coal and just burn clean natural gas. Aren't those plants more efficient anyway?" That's true... until the price of gas goes up again. Remember that $450M? You'll need that for a down payment on your new clean gas-burning plant. This past year I tested such a plant that didn't produce as much power as Bull Run, yet cost $1.4B to construct. Even at that, this high tech combined cycle plant (gas turbines on the top end plus steam turbines on the bottom end) was a bargain compared to wind or solar.

Wind is the way to go... until the things start decapitating bald eagles. Solar panels are the future... until you see the special on how the manufacturers have dumped toxic waste into streams, turned villages into ghost towns, and caused countless birth defects. Oh yeah, the windmill people do that to, making the magnets and wire. Let's harness waves! Dolphins will love playing with the big floaty things, bobbing in the surf at the beach. How about tidal power? I've been to the power plant at the Bay of Fundi. The most spectacular tides in the world can power 13,000 homes. Bull Run can't idle that slowly. It takes more power than that to run the cooling fans at Browns Ferry.

"Out of sight, out of mind," is a common expression because it describes human behavior so well. Let's say we borrow $1.6B and replace Bull Run with the very latest, efficient, natural gas-burning combined cycle plant. Ignore for the moment how much we'd have to raise rates to pay for this swap. Given the higher efficiency and heating value of natural gas, the new plant would consume over 3,500 tons of natural gas per day. That's 142 million cubic feet per day, but it flows underground where we can't see it and worry about it... until the pipeline comes through our yard or it springs a leak and takes out a subdivision. Can it really do that?

There used to be a house here somewhere.

The problem with all these seemingly good ideas for cheap, clean power is that the people suggesting them have no idea what they're talking about, have no grasp on how much power we consume, and have no concept of how many people there are in the world. There are enough people on this planet to simultaneously fill 74,000 football stadiums. They all want power and most of them will complain if it goes off for more than 5 minutes. Oh yeah... It must be cheap too.

Don't worry. This is just a local thing. It can't happen where you live.

Chapter 13. The Problem with Giants

There are many legends that contain giants. As we consider these, it is very important to separate them into two categories: 1) very large humans, 2) mythological creatures. Robert Pershing Wadlow (1918-1940) was 8'11". André René Roussimoff, the beloved Fezzik in *The Princess Bride*, was 7'4". There are, and have been, very large humans. These are not *giants* in the proper sense of the word. This figure shows the mythical Paul Bunyan.

58

If this picture were accurate, Paul would be 15 feet tall, or 2½ times as tall as his little guitar-strumming friend, Dusty. More importantly, Paul would weigh in at over 2700 pounds, compared to Dusty's 175. Paul's thigh would be as big as Dusty. The following chart purports to list ideal weight:

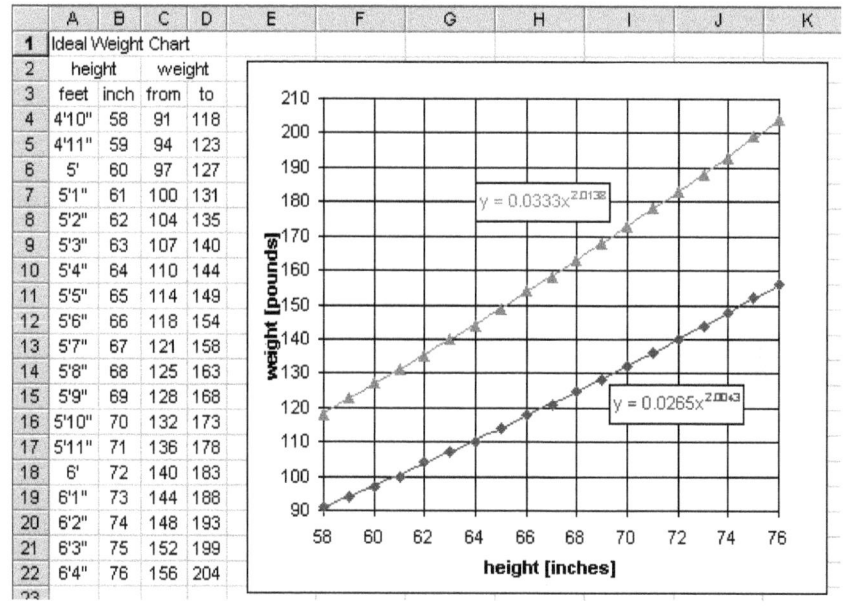

	A	B	C	D
1	Ideal Weight Chart			
2	height		weight	
3	feet	inch	from	to
4	4'10"	58	91	118
5	4'11"	59	94	123
6	5'	60	97	127
7	5'1"	61	100	131
8	5'2"	62	104	135
9	5'3"	63	107	140
10	5'4"	64	110	144
11	5'5"	65	114	149
12	5'6"	66	118	154
13	5'7"	67	121	158
14	5'8"	68	125	163
15	5'9"	69	128	168
16	5'10"	70	132	173
17	5'11"	71	136	178
18	6'	72	140	183
19	6'1"	73	144	188
20	6'2"	74	148	193
21	6'3"	75	152	199
22	6'4"	76	156	204

Chart fits: $y = 0.0333x^{2.0138}$ (upper) and $y = 0.0265x^{2.0043}$ (lower), with height [inches] on the x-axis and weight [pounds] on the y-axis.

The lower and upper curves have been trended. Both trends have an exponent of 2. If you're tall and think this chart is unfair, you're right. Anyone who considers 6'2" and 148 pounds is normal would probably suggest Mary-Kate lay off the sweets. Someone should have paid more attention in Mr. Drucker's geometry class. Weight is proportional to volume and volume is proportional to length cubed:

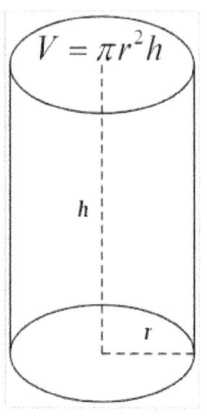

$$V = \pi r^2 h$$

As Paul and Dusty have been depicted with the same proportions, Paul's weight would be 175*(2.5)^3=2734. Not only does h increase by a factor of 2.5, but so does r. That's why V and $W=\rho V$ go up by (2.5)^3. The density (ρ) of humans is very close to that of water and we're assuming Paul is flesh and bone, not styrofoam.

The problem with there being *real* giants having the same proportions as humans, is that muscle strength goes up with the cross-sectional area (i.e., the square), but the weight goes up with the cube. Galileo pointed this out in his book, *Dialogues Concerning Two New Sciences*, published in 1638. Bone strength is also a problem for giants, because the buckling load of a simply supported column (e.g., Paul's femur) is $\pi^2 D^4 E/64L^2$, where D is the diameter, E is the modulus of elasticity, and L is the length.

This size to weight relationship is quite evident in this comparison of the smallest (DikDik) and largest (Eland) antelopes:

It is also why T-Rex is depicted with huge thighs compared to this lizard.

This comparison of human and dinosaur femurs is also stark:

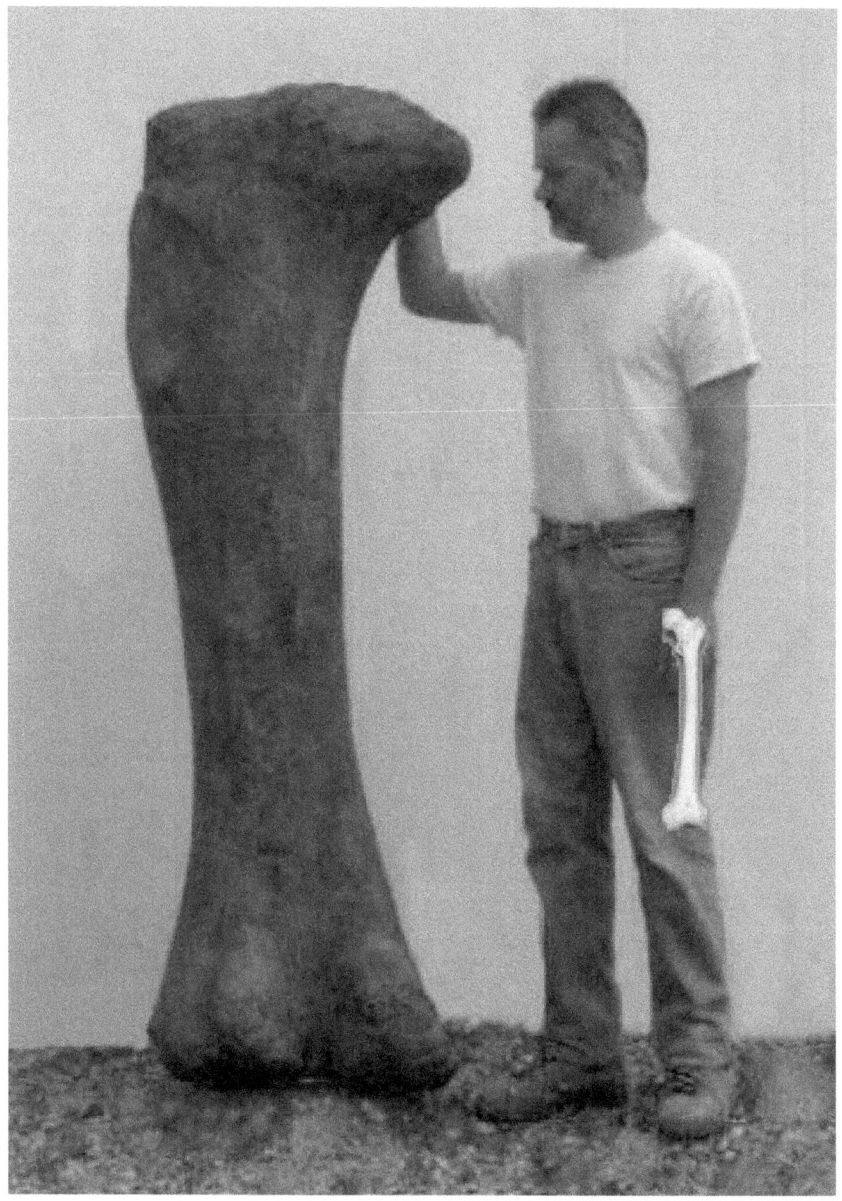

If there were real giants, they wouldn't have the same proportions as humans. This same principle works in the other direction too. As things get smaller, the area (and strength) becomes larger in proportion to the volume (weight), as illustrated by the amazing ant.

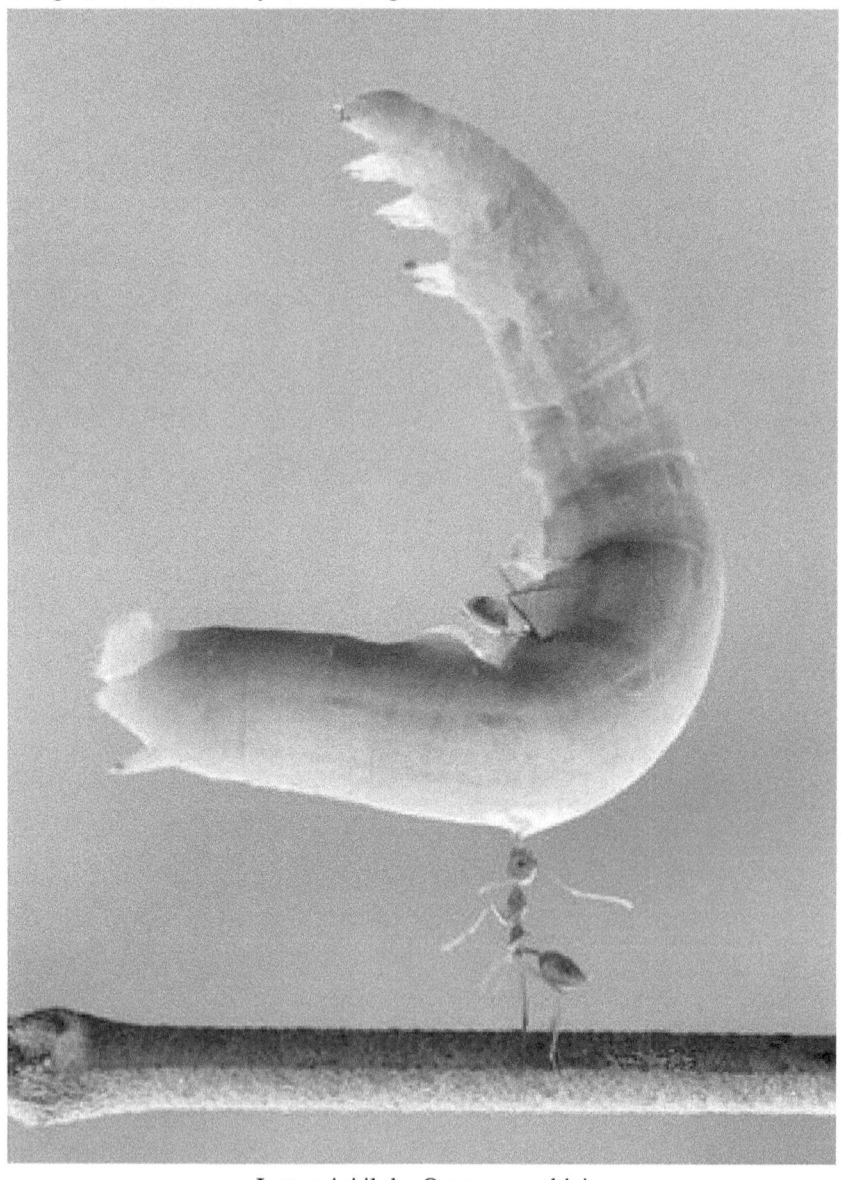

Just wait 'til the Queen sees this!

Chapter 14. Chaos

Since the release of *Jurassic Park*, the terms "chaos," "chaos theory," and "chaotician," have come into common use. In the movie, Dr. Ian Malcolm (Jeff Goldblum) explains chaos theory:

> *A butterfly can flap its wings in Peking and in Central Park you get rain instead of sunshine.*

"Chaos" has become the fashionable term for anything we don't yet understand. "Chaos theory" is the new "magic" invoked with awe to explain the inexplicable, like fairies and wizards, only with pseudoscientific credibility. "Chaoticians" are the new high priests of the chaos cult—practitioners that perform tricks like the witch doctors of old, only on their computers. There's every bit as much chanting and dancing, but the sacrifices are now just cash.

Let's say I present you with a theory that would rightly apply to a chaotic event and get you to agree with the theory under such conditions. I then present you with a data set that defies logical description and get you to agree that the data is chaotic. Finally, I insist that you accept some wacky conclusion. Does this sound far-fetched? Sadly, this scenario is all too common.

There are several fallacies in this approach. We will consider just one: There is no phenomenon in this universe that is chaotic in the strict sense. For instance, random numbers are not chaotic. They're always numbers and they're even finite. In order to be truly chaotic, occasionally these "things" should be colors or sounds or pink elephants reciting bad poetry in Yiddish backwards.

Dr. Malcolm's definition of chaos implies that there's no discernable (or reasonable) connection between cause and effect. Chaos does not arise from simply shuffling reasonable things or constructing an ill-balanced segmented pendulum—one of the countless examples of chaos found on the Web. The pendulum in this Wikipedia article draws an odd path, to be sure, but there's nothing whatsoever chaotic about its behavior.

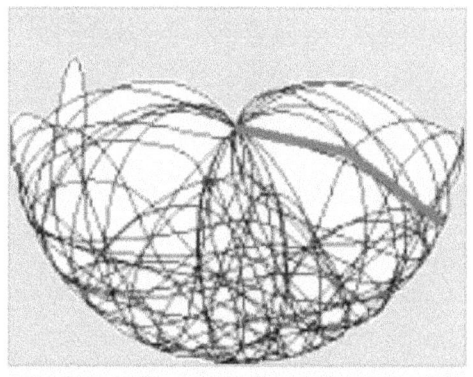

A double rod pendulum

It never even draws in the green shaded area. It always draws in black on white, never chartreuse on mauve. Presumably it conserves energy, as well as both linear and angular momentum. It doesn't jump out of the screen and smack you in the face. It doesn't sing, dance, or play the fiddle. There are an infinite number of things this pendulum will never do. Its behavior is not chaotic. It is ludicrous to associate the following fractal image with chaos, yet the author of this Wikipedia article does just that.

Barnsley fern created using

These are said to be examples of "deterministic chaos." Diabetic treats: Sugar-free sugar. Instant water–don't be caught dry without it! If ever there was an oxymoron, "deterministic chaos" is it. Consider the following list of numbers:

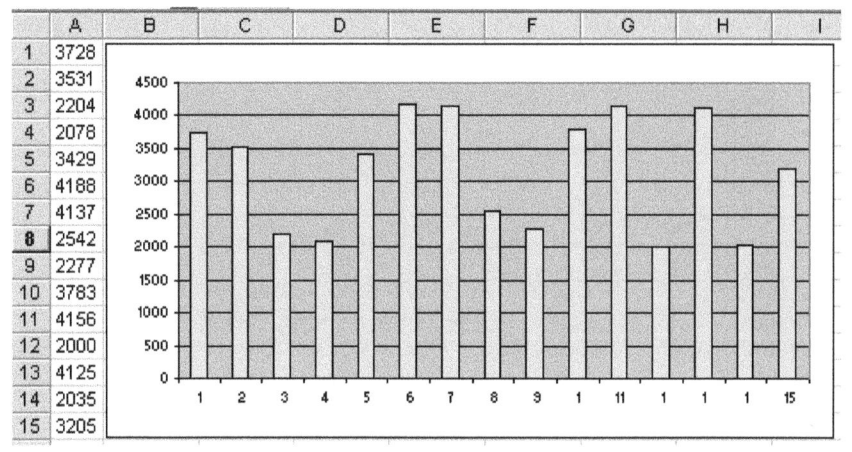

	A	B	C	D	E	F	G	H	I
1	3728								
2	3531								
3	2204								
4	2078								
5	3429								
6	4188								
7	4137								
8	2542								
9	2277								
10	3783								
11	4156								
12	2000								
13	4125								
14	2035								
15	3205								

Are these numbers chaotic? Are they even random? There doesn't seem to be a pattern. In fact, there is no pattern to their values, but that doesn't mean they're random, much less chaotic. They're actually numbers from a very old phone book.

```
Williams Anne R Mrs 210 Diagonal ---------- OR 3-3728
Williams Charles H 185 E 100 N ------------ OR 3-3531
Williams Howard 117 S 600 E --------------- OR 3-2204
Williams Kumen 33 S 500 E ----------------- OR 3-2078
Williams Mary Mrs 317 W 300 S ------------- OR 3-3429
Wilson Agnes 67 N 300 E ------------------- OR 3-4188
Wilson Ellis 768 E 400 S ------------------ OR 3-4137
Winkelmann R C Washington ----------------- OR 3-2542
Winona Robert 528 S 100 W ----------------- OR 3-2277
Wirthlin Jay L 54 W 300 S ----------------- OR 3-3873
Wittwer Clifford E 393 N 400 W ------------ OR 3-4156
Wittwer Erle 70 E 400 S ------------------- OR 3-2000
Wittwer Julius C Santa Clara -------------- OR 3-4125
Wittwer Lester Santa Clara ---------------- OR 3-2035
Wittwer Lorraine Santa Clara -------------- OR 3-3205
```

There is a pattern associated with these numbers. They're alphabetical by owner, but you won't get this from analyzing the numbers. Even if you can't figure out what the pattern is or can't see a relationship between cause and effect, it is the height of arrogance to insist that such doesn't exist.

Unlikely events do occasionally occur, but this doesn't provide proof of impossible events. Many years ago my grandmother taught my wife and I how to play bridge. The very first hand we were ever dealt in the game of bridge contained all of the spades except the 3 and most of the other face cards as well.

This was unusual, but not chaotic. We could not have been dealt 14 spades from a single new deck unless someone had packaged it incorrectly at the factory. Even if we had gotten two rooks, three old maids, and a draw-four, it would not have been chaotic.

In Chapter 9 I told of my personal encounter with a rogue wave. This was an unlikely event. It was scary and strange, but not chaotic. A wave of green jello with red marshmallows conveying a troop of juggling grizzly bears on unicycles singing the French national anthem in Chinese would be chaotic.

Discussions of chaos remind me of watching a magician perform on stage. The show begins with a few simple things to draw you in and direct your attention toward the beautiful assistant and away from the mysterious box. The magician shows you each of the swords and may even bring someone up from the audience to verify that they are real and sharp. The assistant climbs into the box, the swords are thrust through, and everyone gasps.

People who go on and on about chaos are like this magician. They want to pull something over on you. They want you to concede the possibility of impossible events. Even if you saw a woman climb into the box and then the magician pull three monkeys out of the box, it wouldn't be chaos. You just didn't see him make the switch. The monkeys were there all along.

The probability of being dealt four aces in poker is approximately 1 in 54,145. The probability of being dealt 5 aces in a poker hand is precisely zero– the same chance you'd have of leaving the saloon alive after playing such a hand with Billy the Kid, Jesse James, and Butch Cassidy. When someone starts talking about chaos, hold onto your wallet because you're about to get snookered.

Chapter 15. Probability & Chance

Some believe in the Force and others in Luck. The former are drawn to galactic wars and the latter to gambling. There is one thing you can be sure of in a game of chance: the House always wins in the end. The opulent casino and racetrack should be all the proof of this you need. Our exploration of probability and chance will begin with the "one-armed bandit," or slot machine. This is what the old ones look like on the inside:

What you may not know is that the wheels aren't the same. That is, they don't have the same pictures on them. The wheels are designed to give you the impression of having a greater chance of winning than you actually do. These tables show how the pictures work into odds:

picture	reel 1	reel 2	reel 3
bar	3	2	1
cherry	3	3	2
plum	3	4	3
bell	4	4	4
orange	4	4	5
lemon	4	4	6
total	21	21	21

spin	payback	chance	return	
3 bars	1000 to 1	2/3087	65%	
3 cherries	333 to 1	2/1029	65%	
3 plums	167 to 1	4/1029	65%	
3 bells	94 to 1	64/9261	65%	
3 oranges	75 to 1	80/9261	65%	
3 lemons	62 to 1	32/3087	64%	
2 bars	26 to 1	11/441	65%	
2 cherries	14 to 1	1/21	67%	best
2 plums	9 to 1	11/147	67%	
2 melons	6 to 1	16/147	65%	
2 oranges	5 to 1	8/63	63%	
2 lemons	4 to 1	64/441	58%	worst
		average	65%	

There are 3 bars on the first reel, but only 2 on the second and 1 on the third. There are 21 pictures total on each reel. Rather than your chances of getting 3 bars in a row being $(3/21)^3=1/343$, they're only $(3/21)*(2/21)*(1/21)=2/3087$. A payback of 1000:1 for odds of 1:343 sounds too good to be true because it is.

There are numerous articles on the Web listing the chances of being dealt every imaginable hand of cards. Articles on the odds at roulette also abound. There are web sites explaining how you can win the lottery and others explaining why you never will. There's no point repeating them all here; however, we will first consider the chances of being dealt 4 aces. There are some qualifications, for instance, there are 4 players and you are seated to the dealer's left. There are 5 ways of being dealt 4 aces: 1,2,3,4; 1,2,4,5; 1,2,3,5; 1,3,4,5; and 2,3,4,5. The odds are as follows:

4 players - you are to the dealer's left					
you get 4 aces in the first 4 cards dealt to you				
card	player	chance	out of	what
1	1	4	52	that you will get a first ace
1	2	48	51	that player 2 won't get one of your aces
1	3	47	50	that player 3 won't get one of your aces
1	4	46	49	that player 4 won't get one of your aces
2	1	3	48	that you will get a second ace
2	2	45	47	that player 2 won't get one of your aces
2	3	44	46	that player 3 won't get one of your aces
2	4	43	45	that player 4 won't get one of your aces

card	player	chance	out of	what
3	1	2	44	that you will get a third ace
3	2	42	43	that player 2 won't get your fourth ace
3	3	41	42	that player 3 won't get your fourth ace
3	4	40	41	that player 4 won't get your fourth ace
4	1	1	40	that you will get the fourth ace

14,606,122,980,556,800 you now have 4 aces
3,954,242,643,911,240,000,000 1:270,725

you get an ace on card 1, 2, 3, and 5

card	player	chance	out of	what
1	1	4	52	that you will get a first ace
1	2	48	51	that player 2 won't get one of your aces
1	3	47	50	that player 3 won't get one of your aces
1	4	46	49	that player 4 won't get one of your aces
2	1	3	48	that you will get a second ace
2	2	45	47	that player 2 won't get one of your aces
2	3	44	46	that player 3 won't get one of your aces
2	4	43	45	that player 4 won't get one of your aces
3	1	2	44	that you will get a third ace
3	2	42	43	that player 2 won't get your fourth ace
3	3	41	42	that player 3 won't get your fourth ace
3	4	40	41	that player 4 won't get your fourth ace
4	1	39	40	that you won't get the fourth ace
4	2	38	39	that player 2 won't get your fourth ace
4	3	37	38	that player 3 won't get your fourth ace
4	4	36	37	that player 4 won't get your fourth ace
5	1	1	36	that you will get the fourth ace

28,832,837,310,570,700,000,000 you now have 4 aces
7,805,769,880,904,240,000,000,000 1:270,725

you get an ace on card 1, 2, 4, and 5

card	player	chance	out of	what
1	1	4	52	that you will get a first ace
1	2	48	51	that player 2 won't get one of your aces
1	3	47	50	that player 3 won't get one of your aces
1	4	46	49	that player 4 won't get one of your aces
2	1	3	48	that you will get a second ace
2	2	45	47	that player 2 won't get one of your aces
2	3	44	46	that player 3 won't get one of your aces
2	4	43	45	that player 4 won't get one of your aces

card	player	chance	out of	what
3	1	42	44	that you won't get a third ace
3	2	41	43	that player 2 won't get one of your aces
3	3	40	42	that player 3 won't get one of your aces
3	4	39	41	that player 4 won't get one of your aces
4	1	2	40	that you will get the third ace
4	2	38	39	that player 2 won't get your fourth ace
4	3	37	38	that player 3 won't get your fourth ace
4	4	36	37	that player 4 won't get your fourth ace
5	1	1	36	that you will get the fourth ace

28,832,837,310,570,700,000,000 you now have 4 aces
7,805,769,880,904,240,000,000,000,000 1:270,725

you get an ace on card 1, 3, 4, and 5

card	player	chance	out of	what
1	1	4	52	that you will get a first ace
1	2	48	51	that player 2 won't get one of your aces
1	3	47	50	that player 3 won't get one of your aces
1	4	46	49	that player 4 won't get one of your aces
2	1	45	48	that you won't get a second ace
2	2	44	47	that player 2 won't get one of your aces
2	3	43	46	that player 3 won't get one of your aces
2	4	42	45	that player 4 won't get one of your aces
3	1	3	44	that you will get a second ace
3	2	41	43	that player 2 won't get one of your aces
3	3	40	42	that player 3 won't get one of your aces
3	4	39	41	that player 4 won't get one of your aces
4	1	2	40	that you will get the third ace
4	2	38	39	that player 2 won't get your fourth ace
4	3	37	38	that player 3 won't get your fourth ace
4	4	36	37	that player 4 won't get your fourth ace
5	1	1	36	that you will get the fourth ace

28,832,837,310,570,700,000,000 you now have 4 aces
7,805,769,880,904,240,000,000,000,000 1:270,725

you get an ace on card 2, 3, 4, and 5

card	player	chance	out of	what
1	1	48	52	that you won't get a first ace
1	2	47	51	that player 2 won't get one of your aces
1	3	46	50	that player 3 won't get one of your aces
1	4	45	49	that player 4 won't get one of your aces

2	1	4	48	that you will get a first ace
2	2	44	47	that player 2 won't get one of your aces
2	3	43	46	that player 3 won't get one of your aces
2	4	42	45	that player 4 won't get one of your aces
3	1	3	44	that you will get a second ace
3	2	41	43	that player 2 won't get one of your aces
3	3	40	42	that player 3 won't get one of your aces
3	4	39	41	that player 4 won't get one of your aces
4	1	2	40	that you will get the third ace
4	2	38	39	that player 2 won't get your fourth ace
4	3	37	38	that player 3 won't get your fourth ace
4	4	36	37	that player 4 won't get your fourth ace
5	1	1	36	that you will get the fourth ace
		28,832,837,310,570,700,000,000		you now have 4 aces
	7,805,769,880,904,240,000,000,000,000		1:270,725	
			5	ways to get 4 aces
			260,725	1:52,145

Given these conditions, your chances of being dealt 4 aces when seated to the left of the dealer is 1 in 52,145.

Chapter 16. Extrapolating into the Future

The next question we will consider is where do those 100-year and 500-year projections come from if we've only been collecting data for the past 25 years? The simple answer may be that someone made them up, but there is a scientific way of coming up with these projections. Explanation of this process starts with a special kind of graph paper–something you may never have seen.

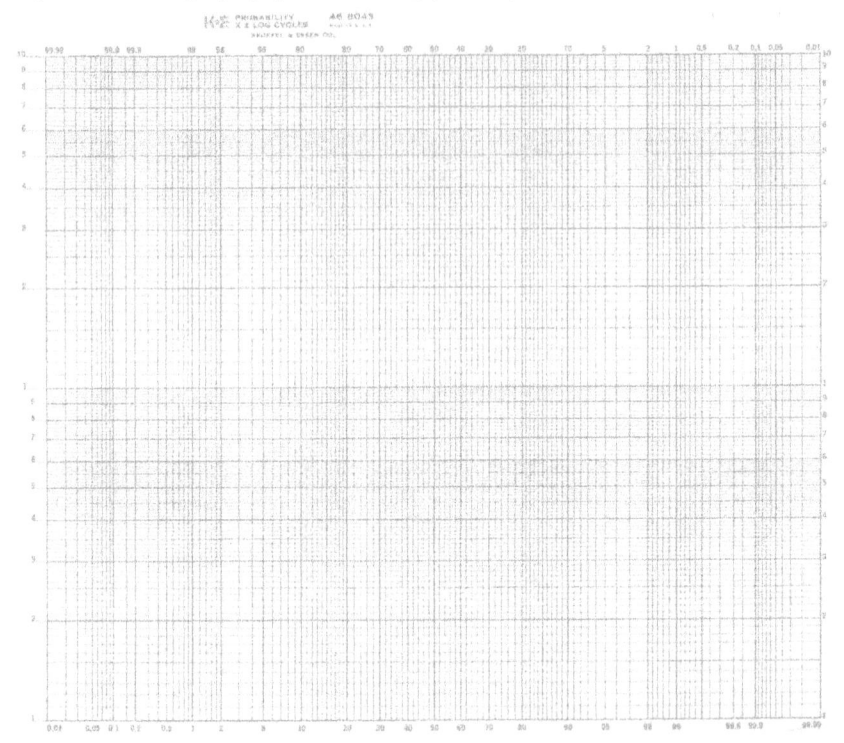

This is a close-up of the scale at the top, split down the middle, and wrapped back around:

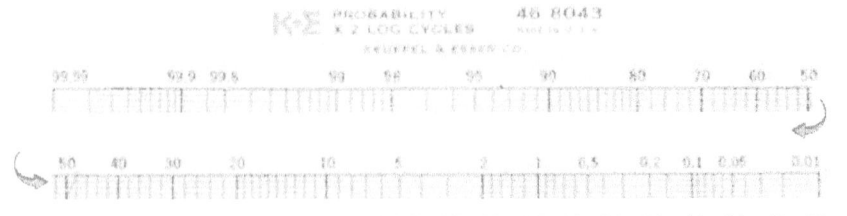

The numbers are: 99.99, 99.9, 99.8, 99, 98, 95, 90, 80, 70, 60, 50, 40, 30, 20, 10, 5, 2, 1, 0.5, 0.2, 0.1, 0.05, and 0.01. These are percentages. The center of the scale is at 50%. The right side is a mirror image of the left side and the divisions on the right side are equal to 100% minus the divisions on the left.

This is a *probability* scale. It is quite different from a log scale. This scale and paper was common at one time, but is surprisingly scarce on the Web. This is likely due to the fact that Excel® won't draw a probability axis.

There is a mathematical equivalent that can be used instead of this scale, but it doesn't convey the same sense of value as this scale does. The distance between 99.99 and 99.9 is about the same as between 99.8 and 99, or 80 and 60, or 40 and 20. It's exactly the same as the distance between 0.1 and 0.01 because the right side is a mirror image of the left.

The usefulness of this scale becomes apparent when you realize that a cumulative normal distribution (i.e., the area under a bell-shaped curve) plots as a straight line against this axis. To illustrate this, consider the following fabricated list of 36,721 IQ scores:

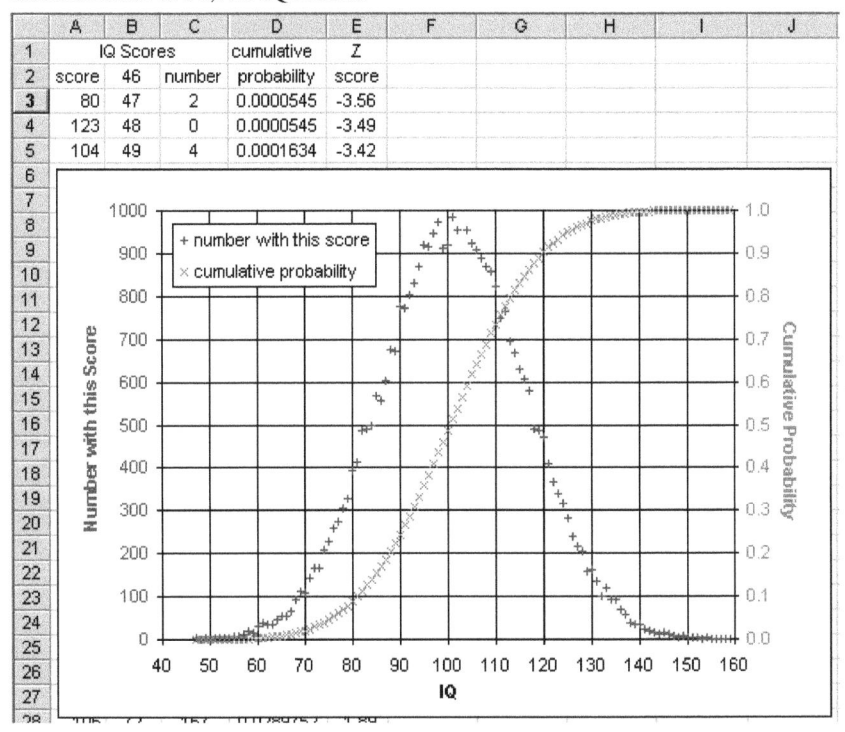

	A	B	C	D	E
1	IQ Scores			cumulative	Z
2	score	46	number	probability	score
3	80	47	2	0.0000545	-3.56
4	123	48	0	0.0000545	-3.49
5	104	49	4	0.0001634	-3.42

The blue +s indicate the number of persons with each score and the red xs indicate the cumulative probability. The blue +s have the familiar bell shape and the red xs are the area under the blue curve normalized to 1. This is what the red xs look like plotted on a probability scale:

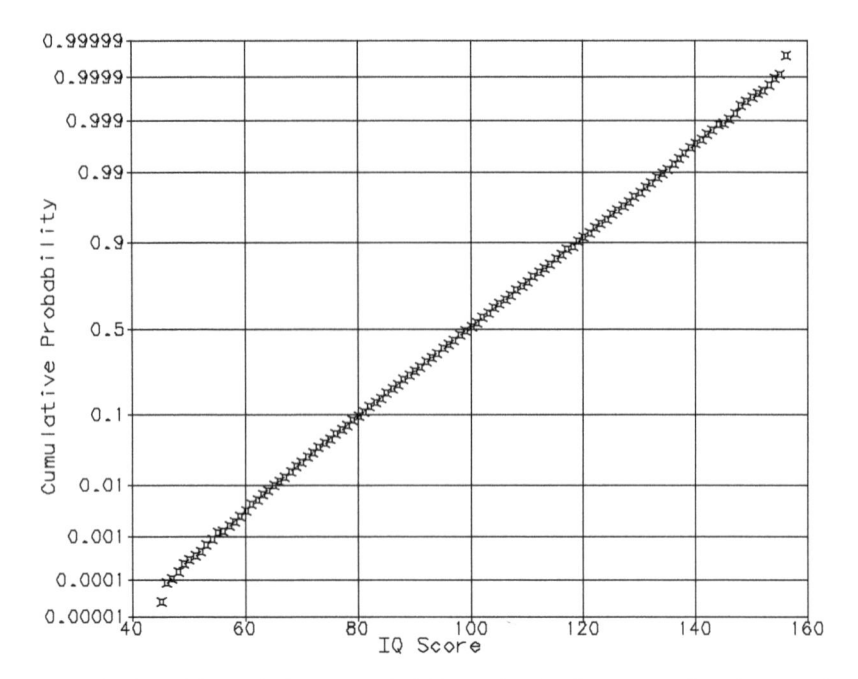

While Excel® won't draw a probability axis, the Z-score (or Z-value) can be used instead. Z is given by the following equation:

$$Z_i = \frac{X_i - \overline{X}}{\sigma} \qquad (16.1)$$

In this equation X_i is the usual score (or value), $\overline{X}$ is the mean (i.e., average), and σ is the standard deviation. The Z values for this data are listed in column E of the spreadsheet above and plotted as the Y-axis in the following figure:

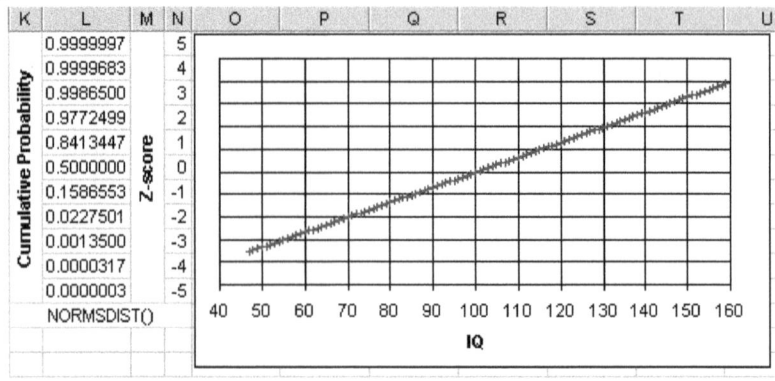

Although the Y-axis numbering doesn't show 0.00001 to 0.99999 as in the preceding graph, the shape is exactly the same. While Excel® won't draw a probability axis, it does have functions that facilitate working with probability. These include: NORMDIST(), NORMINV(), NORMSDIST(), NORMSINV(). The integrated help feature describes these functions. Excel® also has more versatile probability functions, including: BETADIST(), BETAINV(), CHIDIST(), CHIINV(). The following table shows the correspondence between probability and Z-value as calculated using the NORMSINV() function:

	A	B
1	probability	NORMSINV
2	0.000001	-4.77
3	0.00001	-4.27
4	0.0001	-3.72
5	0.001	-3.09
6	0.01	-2.33
7	0.1	-1.28
8	0.5	0.00
9	0.9	1.28
10	0.99	2.33
11	0.999	3.09
12	0.9999	3.72
13	0.99999	4.27
14	0.999999	4.77

Comparing this table with the left side in the preceding figure shows that convenient values of the Z-value don't correspond to convenient values of probability.

Consider the following 21.5 years of historical data for the Mississippi River at Memphis:

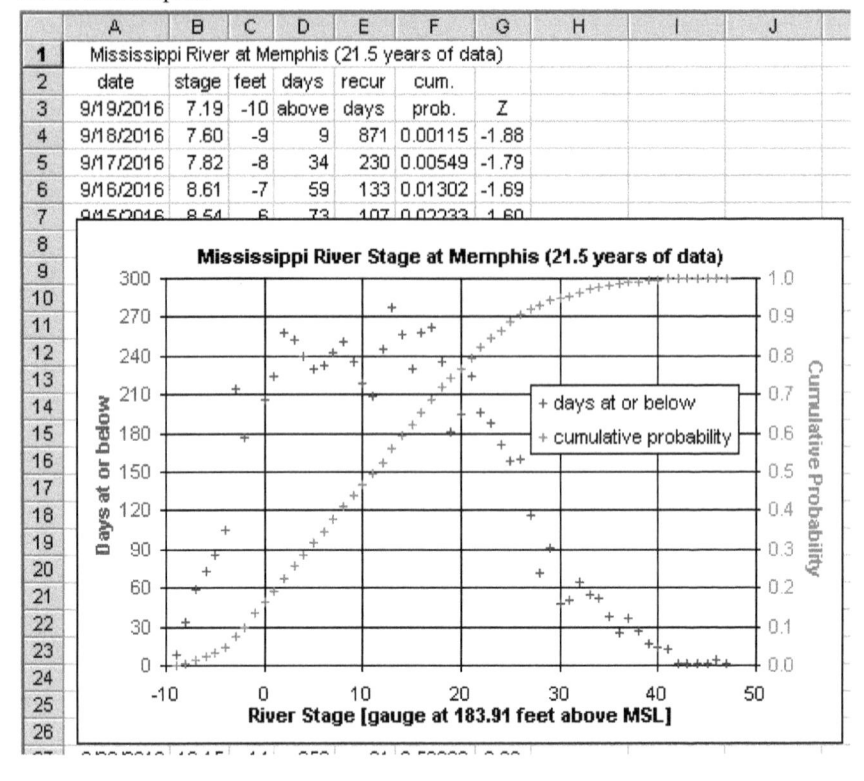

	A	B	C	D	E	F	G	H	I	J
1	Mississippi River at Memphis (21.5 years of data)									
2	date	stage	feet	days	recur	cum.				
3	9/19/2016	7.19	-10	above	days	prob.	Z			
4	9/18/2016	7.60	-9	9	871	0.00115	-1.88			
5	9/17/2016	7.82	-8	34	230	0.00549	-1.79			
6	9/16/2016	8.61	-7	59	133	0.01302	-1.69			
7	9/15/2016	8.54	-6	73	107	0.02233	-1.60			

Mississippi River Stage at Memphis (21.5 years of data)

+ days at or below
+ cumulative probability

x-axis: River Stage [gauge at 183.91 feet above MSL]
left y-axis: Days at or below
right y-axis: Cumulative Probability

The cumulative probabilities are then plotted against the Z-values:

L	M	N	O	P	Q	R	S	T
	projection							
ft	Z	P	years					
48.98	3.555	0.99981	50					
50.88	3.734	0.99991	100					
52.70	3.905	0.99995	200					
55.01	4.121	0.99998	500					

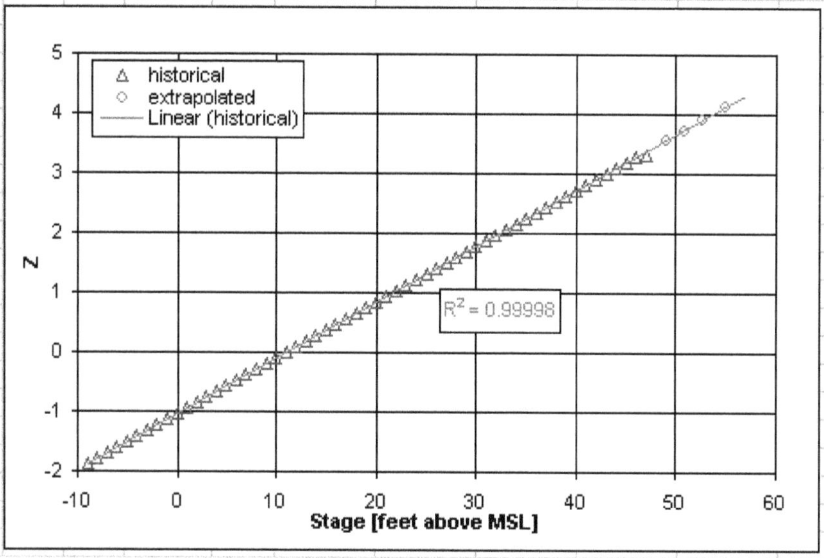

[Chart: Z vs. Stage [feet above MSL], with historical (△), extrapolated (○), and Linear (historical) series; $R^2 = 0.99998$]

The relationship between stage and Z can be used to extrapolate out to 50, 100, 200, and 500 years, as shown in the table and figure above. In this case I used the solver to find the stage (48.98, 50.88, 52.70, and 55.01) that corresponded to certain values of P (1:50, 1:100, 1:200, and 1:500 years). This same methodology can be applied to all sorts of forecasting. This process provides an estimate, not a guarantee. "Only time will tell," applies here.

Chapter 17. Monte Carlo Simulations

Monte Carlo simulations derive their name from the famous casino. These attempt to simulate processes through the use of random numbers and a large number of cases so that statistics can be compiled on the outcomes. The first thing we need in order to build a Monte Carlo simulation is a random number generator. The rand() function in C and randbetween() function Visual Basic® produce uniformly-distributed random numbers, that is, random numbers that have a flat probability distribution. What we need is normally-distributed random numbers that have a bell-shaped probability distribution.

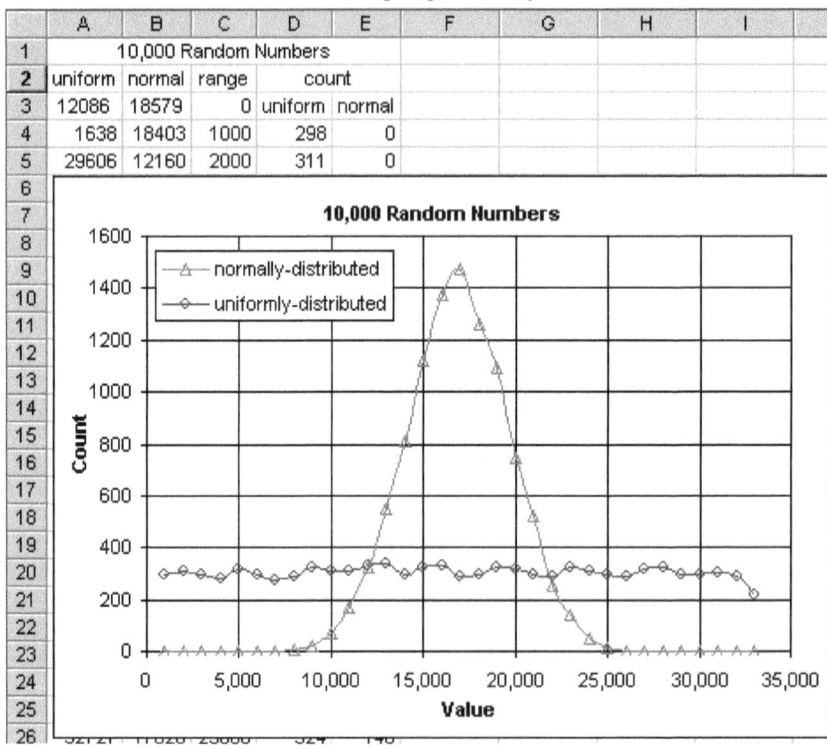

	A	B	C	D	E	F	G	H	I
1	10,000 Random Numbers								
2	uniform	normal	range	count					
3	12086	18579	0	uniform	normal				
4	1638	18403	1000	298	0				
5	29606	12160	2000	311	0				

10,000 Random Numbers

The simplest way to produce normally-distributed integers from uniformly-distributed ones is the following formula:

```
int urand()
{
int i,r;
for(r=i=0;i<12;i++)
   r+=rand();
return(r/12);
}
```

The rand() function, as defined in stdlib.h, returns uniformly-distributed unsigned 15-bit integers. The integers *i* and *r* are 32-bit signed integers and will not overflow in this case. For real numbers a slight modification of this formula can be used to create normally-distributed numbers having a mean of 0 and a standard deviation of 1:

```
double randnorm()
{
int i;
double r;
for(r=i=0;i<12;i++)
   r+=rand()/32767.;
return((r-6.)/6.);
}
```

The function above can be used to create normally-distributed numbers having a mean of *a* and s standard deviation of *s*:

```
double randist(double a,double s)
{
return(a+6.*s*randnorm());
}
```

The following function will produce signed 16-bit uniformly-distributed random integers:

```
int irand()
{
if(rand()%2)
   return(-rand());
return(rand());
}
```

The following function will produce 32-bit uniformly-distributed random integers:

```
int jrand()
{
union{short s[2];int i;}u;
u.s[0]=irand();
u.s[1]=irand();
return(u.i);
}
```

While there have been many papers written on the deficiencies of various random number generating algorithms, such as the one used in rand(), these are rarely the weak links in a Monte Carlo simulation. For our first simulation, consider a battle between an orc (challenge rating 4) and a troll (challenge rating 5) or a hydra ((challenge rating 8). The simplest calculation would be following:

```
#include <stdio.h>
#include <stdlib.h>
int urand()
{
```

```
  int i,u;
  for(u=i=0;i<12;i++)
    u+=rand();
  return(u);
}
#define   orc()   (4*urand())
#define   troll() (5*urand())
#define   hydra() (8*urand())
int main(int argc,char**argv,char**envp)
{
  int i,j,k,l;
  for(i=j=k=0;i<10000;i++)
    {
    l=orc()-troll();
    if(l>0)
      j++;
    else if(l<0)
      k++;
    }
  printf("orc vs. troll: win=%i, lose=%i,
    tie=%i\n",j,k,i-j-k);
  for(i=j=k=0;i<10000;i++)
    {
    l=orc()-hydra();
    if(l>0)
      j++;
    else if(l<0)
      k++;
    }
  printf("orc vs. hydra: win=%i, lose=%i,
    tie=%i\n",j,k,i-j-k);
  return(0);
}
```

This produces the following result:

```
orc vs. troll: win=1734, lose=8266, tie=0
orc vs. hydra: win=20, lose=9980, tie=0
```

The probabilities are shown in this next figure:

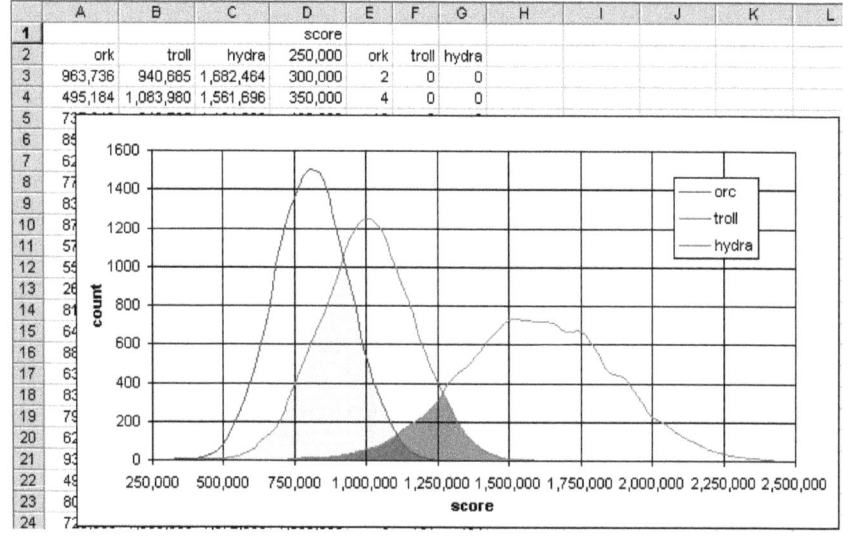

	A	B	C	D	E	F	G	H	I	J	K	L
1				score								
2	ork	troll	hydra	250,000	ork	troll	hydra					
3	963,736	940,685	1,682,464	300,000	2	0	0					
4	495,184	1,083,980	1,561,696	350,000	4	0	0					

The area under each curve is the same (i.e., 10,000 battles). The orc scores lower more often (i.e., the blue curve is to the left and taller). The hydra scores higher less often (i.e., the red curve is to the right and shorter). The troll scores between these two. The overlapping regions have been shaded. The magenta plus green regions are where the troll defeats the hydra. The yellow plus green regions are where the orc defeats the troll. The green region is where the orc defeats the hydra.

Consider this actual manufacturing problem that one automaker had with production of a 6-cylinder engine. The cylinder inside diameter was 3.73820 inches. The accuracy of their current cylinder boring machines was ±0.00061 inch. The desired piston clearance was 0.00220 inch, making the nominal piston diameter 3.73600 inches. The accuracy of their current piston lathes was ±0.00058 inch. They planned to make 1,000,000 engines in their first run. We can approach this problem with a Monte Carlo simulation. The results are shown in the table and figure below:

	A	B	C	D	E	F	G	H	I
1	cylinder	piston	clearance		6,000,000 units		1,023,169	too small	
2	3.73820	3.73600	0.00220		clearance	count	1,023,166	too large	
3	0.00061	0.00058	0.00080		-0.00193	1	2,046,335	outside	
4	3.74023	3.73779	0.00436	max	-0.00183	1	34.1%	bad	
5	3.73620	3.73397	-0.00056	min	-0.00182	1	285,883	too small	
6	3.73788	3.73605	0.00183		-0.00174	1	285,853	too large	
7	3.73759	3.73546	0.00213		-0.00173	1	571,736	outside	
8	3.73811	3.73657	0.00155		-0.00171	1	9.5%	bad	

The target range on clearance was ±0.00080 inch, represented by the vertical dotted red lines. At this level, 1,023,169 piston/cylinder combinations would have too little clearance and another 1,023,166 would have too much clearance, with 34.1% of the total out of spec.

They next considered extending the range of clearance to ±0.00140 inch, represented by the vertical dashed brown lines. At this level, 285,883 piston/cylinder combinations would have too little clearance and another 285,853 would have too much clearance, with only 9.5% out of spec.

The problem is further complicated by the fact that this is a 6-cylinder engine. We will next use this same simulation to determine the number of engines that might have 1, 2, 3, or more piston/cylinder combinations out of spec. The following table shows the same results by engine:

	engines out of spec.			
cyl	0.00080		0.00140	
0	79,363	8%	543,528	54%
1	249,959	25%	348,716	35%
2	329,366	33%	93,317	9%
3	230,041	23%	13,360	1%
4	90,511	9%	1,045	0%
5	19,084	2%	34	0%
6	1676	0%	0	0%
tot	1,000,000	100%	1,000,000	100%

At the original target clearance, only 79,363 engines (8%) would have no cylinders out of spec and 1676 would have all of them out of spec. At the extended clearance, 543,528 engines (54%) would have no cylinders out of spec and none with all of them out of spec. Simply changing the clearance criteria doesn't solve any problems. The next issue to consider is the possible decrease in engine life, as these would be under warranty. We must link clearance to longevity and then run the simulation again.

For the purposes of our simulation, we will assume that an engine with the target clearance has an expected life of 160,000 miles. Doubling the clearance cuts the engine life in half. Halving the clearance cuts the engine life to one-fourth. This relationship is shown in the following table and figure:

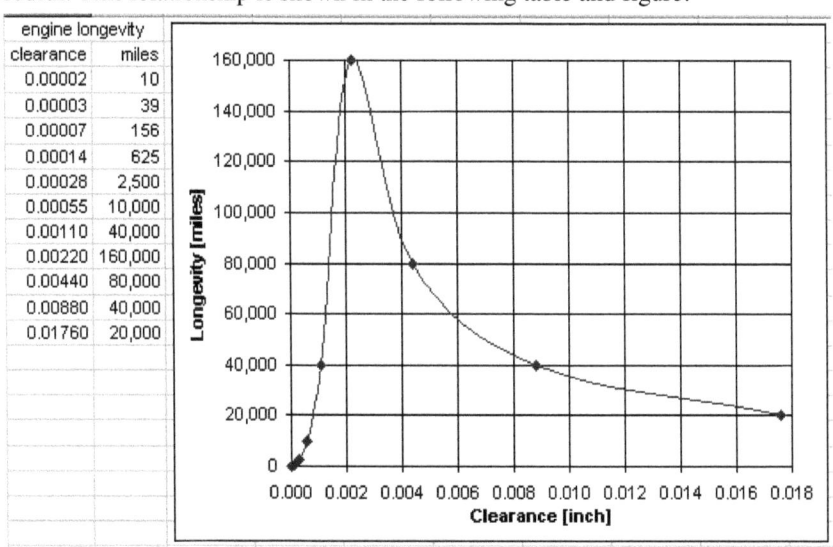

engine longevity	
clearance	miles
0.00002	10
0.00003	39
0.00007	156
0.00014	625
0.00028	2,500
0.00055	10,000
0.00110	40,000
0.00220	160,000
0.00440	80,000
0.00880	40,000
0.01760	20,000

As this is a 6-cylinder engine, the calculations will be performed for each cylinder and the shortest cylinder lifetime will be assigned to the engine. The grim news for this auto manufacturer is evident in this next figure:

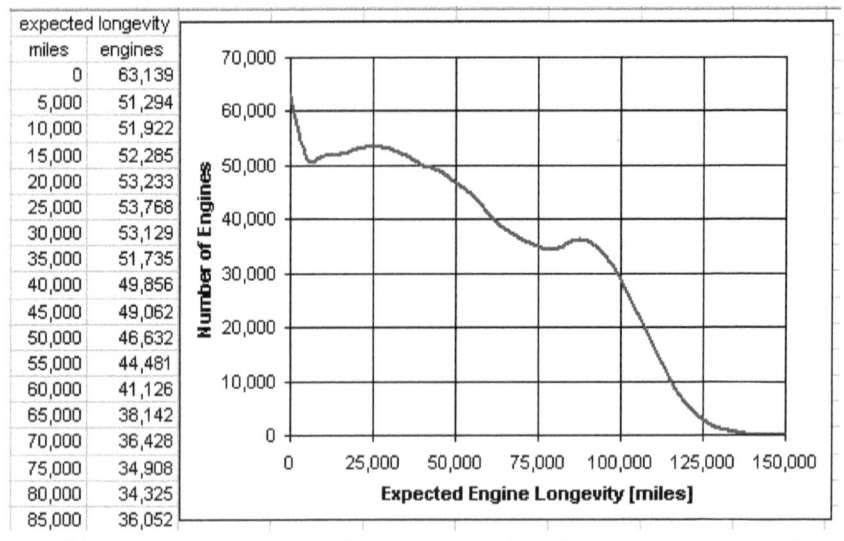

expected longevity	
miles	engines
0	63,139
5,000	51,294
10,000	51,922
15,000	52,285
20,000	53,233
25,000	53,768
30,000	53,129
35,000	51,735
40,000	49,856
45,000	49,062
50,000	46,632
55,000	44,481
60,000	41,126
65,000	38,142
70,000	36,428
75,000	34,908
80,000	34,325
85,000	36,052

This analysis shows that 63,139 of the 1,000,000 engines wouldn't have made it off the lot. As for the warranty issue, 911,008 of the engines (91%) wouldn't have lasted 100,000 miles. None of the engines were expected to last 160,000 miles. This forced the manufacturer to begin building the engines at a newer facility, close the first facility they had hoped to continue using, buy additional modern equipment, and deliver the cars to the dealers at a reduced pace. The program that produced these calculations is provided in Appendix A and is included in the on-line archive.

The last Monte Carlo simulation we will consider is that of a heat exchanger. This problem is based on ASME PTC 12.5, "Single Phase Heat Exchangers," Example K-1. The variables are:

$$\dot{m}_H = 6.00 \pm 0.194$$
$$\dot{m}_C = 10.22 \pm 0.325$$
$$C_{PH} = 0.998 \pm 0.010$$
$$C_{PC} = 0.998 \pm 0.010$$
(17.1-4)

$$T_{HI} = 95.27 \pm 0.27$$
$$T_{HO} = 85.53 \pm 0.43$$
$$T_{CI} = 79.91 \pm 0.21$$
$$T_{CO} = 85.75 \pm 0.66$$
(17.5-8)

The governing equations are:

$$Q_H = \dot{m}_H C_{PH}(T_{HI} - T_{HO})$$
$$Q_C = \dot{m}_C C_{PC}(T_{CO} - T_{CI})$$
$$Q_{AVG} = \frac{Q_H + Q_C}{2} = U \cdot A \cdot LMTD \qquad (17.9\text{-}12)$$
$$LMTD = \frac{\Delta T_1 - \Delta T_2}{\ln\left(\frac{\Delta T_1}{\Delta T_2}\right)}$$

The results of the simulation for 10,000 cases are shown in this next figure:

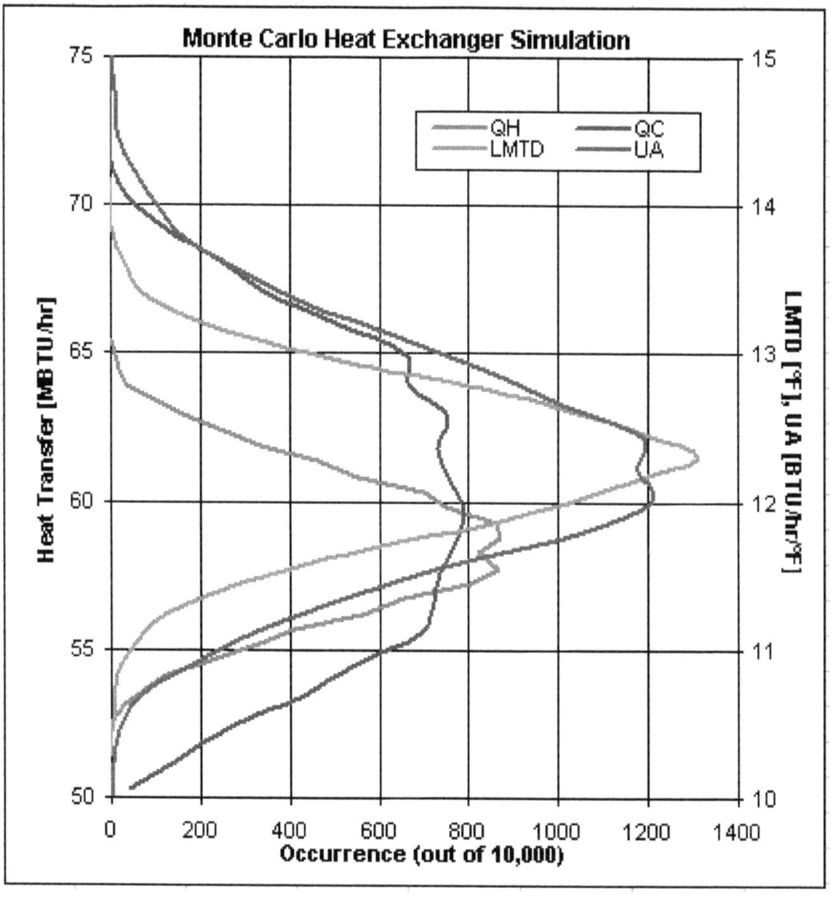
Monte Carlo Heat Exchanger Simulation
QH
QC
LMTD
UA
75
70
65
60
55
50
15
14
13
12
11
10
0
200
400
600
800
1000
1200
1400
Occurrence (out of 10,000)
Heat Transfer [MBTU/hr]
LMTD [°F], UA [BTU/hr/°F]

This graph is rotated, as Excel® will do 2 Y-axes, but not 2 X-axes. The spreadsheet is included in the on-line archive.

Chapter 18. Time Series Simulation

We have already presented a time series simulation in Chapter 1 using Equation 1.18. Here we will use a simulation to evaluate the cost-effectiveness of producing electrical power with the Tellico Dam on the Little Tennessee River. Construction of this dam was quite contentious and resulted in many heated arguments, one of which was whether or not the dam should be used for power production. Construction began in 1967 with operation commencing on 11/29/1979. There is no provision for power production.

There are several physical characteristics needed in order to build this simulation. The Tellico watershed is 655 square miles. The maximum elevation is 817.5 feet above mean sea level (MSL). The target minimum water surface elevation is 807 feet above MSL. The dam is 129 feet high. The elevation to area and volume relationships are shown in the following figure:

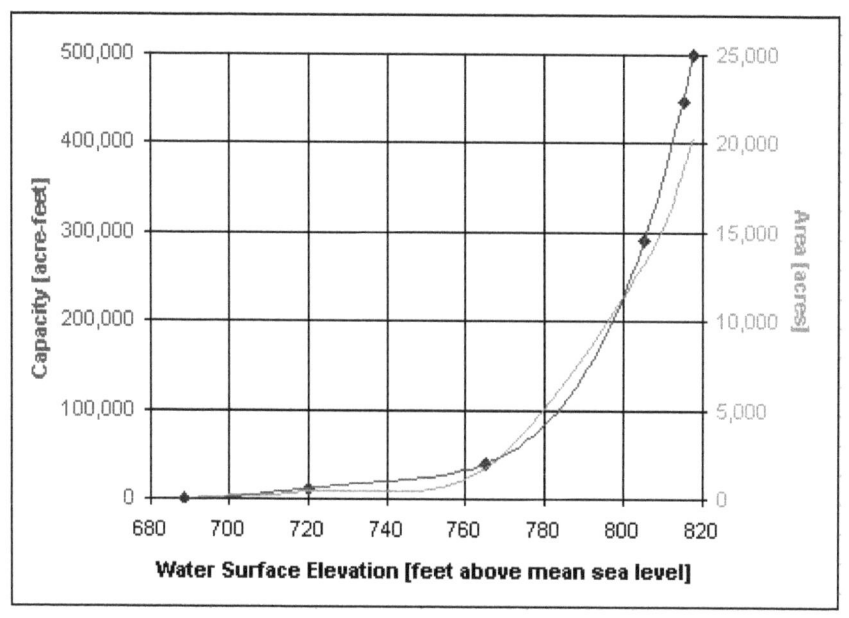

The area is determined from surveys and contours and the volume is determined by integrating the area with respect to depth.

$$V(d) = \int_0^d A(d)dz \qquad (18.1)$$

First the area is fitted with a curve, and then this is integrated analytically:

area=((0.02417924142*d-2.399280326)*d+65.02552476)*d (18.2)
volume=((0.02417924142/4*d-2.399280326/3)*d+65.02552476/2)*d*d (18.3)

The inflow will be based on precipitation as reported by the nearest airport (McGhee-Tyson). A delay always accompanies runoff. This will be approximated by assuming that one-seventh arrives each day following a rain so that it will be spread out over a week. This is mathematically equivalent to a seven day rolling average.

It is also necessary to define a minimum and maximum discharge flow so as to design the turbine and generator. Based on 20 years of rainfall data, the maximum flow will be set at 5000 cubic feet per second (cfs). The minimum flow (when the water surface elevation is at or above the minimum) has considerable impact on the simulation results. The most interesting graphs are obtained with a minimum flow of 2000 cfs. The spreadsheet is included in the archive so that you can experiment with other values.

The flow is not continuously variable. In this case the flow will be increased or decreased in steps of 500 cfs. This parameter can be changed to see it's impact on the results, which is relatively small.

This would qualify as a low head dam; furthermore, the tailwater is controlled by Fort Loudoun Dam, less than 2 miles away. While the turbines in traditional dams may achieve an efficiency of over 90%, low head turbines are considerably less efficient. A value of 60% is used in the simulation. The minimum head is also an important factor. In this case 5 feet is used. The value of power produced is based on a typical commercial rate of $43/MWh (megawatt-hour). This is how the simulation is set up:

	A	B	C	D	E	F	G	H	I
1				Tellico Dam Simulation					
2	date	prcp	watershed	655	miles²	60%	generator		
3	m/d/yyyy	inch	max elev	817.5	ft MSL	5.0	min head		
4	11/20/1979	0.00	min elev	807.1	ft MSL	$43.00	per MWh		
5	11/21/1979	0.00	tailwater	807.0	ft MSL				
6	11/22/1979	0.00	max flow	5000	ft³/s				
7	11/23/1979	0.00	min flow	2000	ft³/s				
8	11/24/1979	1.18	filling begins	11/29/79	acre	elev			
9	11/25/1979	0.16	acre		feet	feet			
10	11/26/1979	0.51	feet		of	above			
11	11/27/1979	0.00	of	discharge	storage	MSL	power	generation value	
12	11/28/1979	0.04	inflow	ft³/s	0	688.50	kW	daily	cumulative
13	11/29/1979	0.01	9,432	0	9,432	712.02	0	$0	$0
14	11/30/1979	0.00	9,482	0	18,914	737.85	0	$0	$0
15	12/1/1979	0.00	9,482	0	28,396	756.80	0	$0	$0
16	12/2/1979	0.00	3,593	0	31,989	760.10	0	$0	$0
17	12/3/1979	0.00	2,795	0	34,784	762.20	0	$0	$0
18	12/4/1979	0.00	250	0	35,033	762.37	0	$0	$0
19	12/5/1979	0.00	250	0	35,283	762.54	0	$0	$0
20	12/6/1979	0.00	50	0	35,333	762.58	0	$0	$0
21	12/7/1979	0.00	0	0	35,333	762.58	0	$0	$0
22	12/8/1979	0.00	0	0	35,333	762.58	0	$0	$0
23	12/9/1979	0.00	0	0	35,333	762.58	0	$0	$0
24	12/10/1979	0.00	0	0	35,333	762.58	0	$0	$0
25	12/11/1979	0.00	0	0	35,333	762.58	0	$0	$0
26	12/12/1979	0.00	0	0	35,333	762.58	0	$0	$0
27	12/13/1979	0.01	0	0	35,333	762.58	0	$0	$0
28	12/14/1979	0.59	50	0	35,382	762.61	0	$0	$0
29	12/15/1979	0.00	2,994	0	38,377	764.49	0	$0	$0
30	12/16/1979	0.00	2,994	0	41,371	766.16	0	$0	$0
31	12/17/1979	0.00	2,994	0	44,365	767.65	0	$0	$0
32	12/18/1979	0.00	2,994	0	47,360	769.01	0	$0	$0
33	12/19/1979	0.00	2,994	0	50,354	770.26	0	$0	$0
34	12/20/1979	0.00	2,994	0	53,348	771.42	0	$0	$0

The predicted daily water surface elevations along with the one-year rolling average is shown in this next figure:

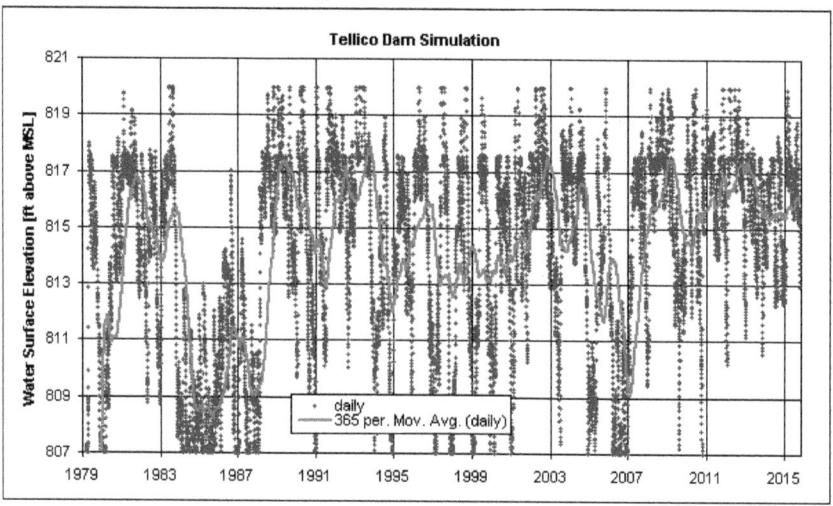

[Figure: Tellico Dam Simulation — Water Surface Elevation (ft above MSL) vs. year, 1979–2015, showing daily values and 365-per. Mov. Avg. (daily), ranging roughly 807–821 ft.]

This next figure shows the expected daily power output and cumulative revenue generated:

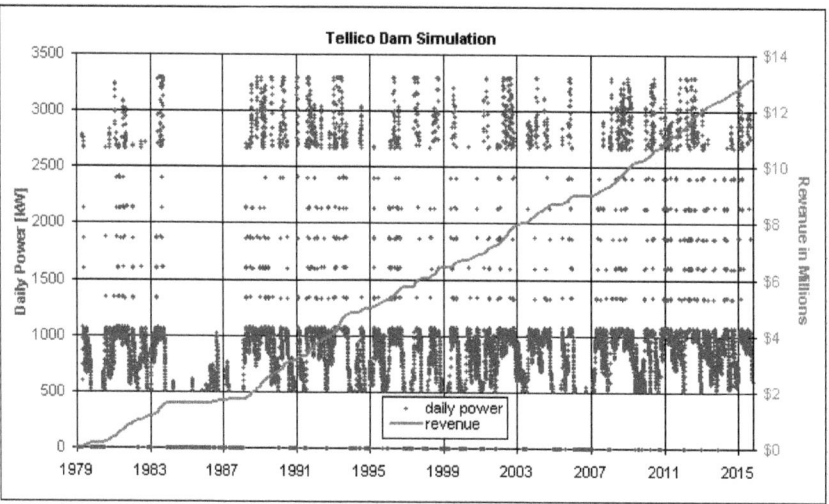

[Figure: Tellico Dam Simulation — Daily Power [kW] and Revenue in Millions vs. year, 1979–2015, showing daily power and cumulative revenue.]

The power output is between 500 and 1000 kW most of the year with 2500 to 3500 kW during the rainy season. The revenue after almost 37 years of operation is not expected to reach $14M, which is nowhere near enough to pay for the dam, let alone a turbine and generator. This dam was clearly built to create lakefront property.

Chapter 19. Solar Power

Solar power production is often quite disappointing due to rosy predictions that don't consider the realities of the sky. When evaluating the performance of solar energy collectors it is often assumed that the Direct Normal Irradiance (DNI) only varies with the position of the sun and is spatially invariant; but this is not the case, even in locations known for their clear skies.

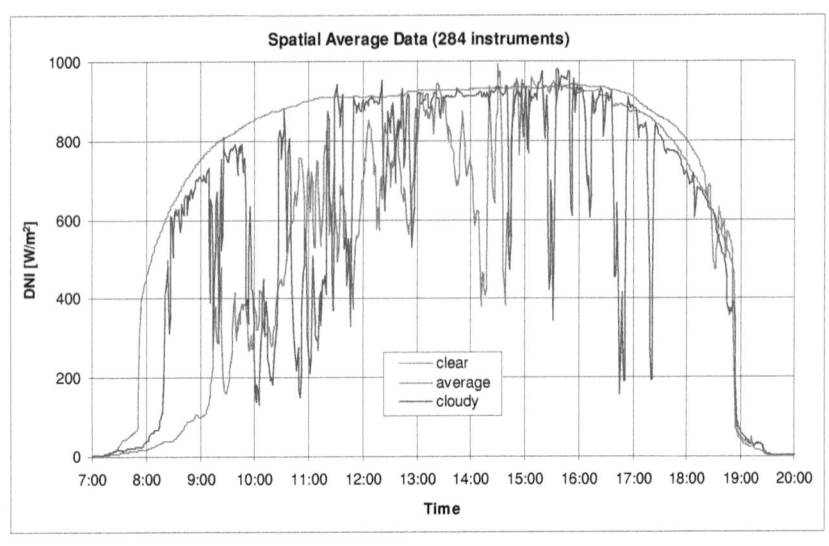

Spatial Average Data (284 instruments)

Three days have been selected from an extensive database: clear, average, and cloudy. As these three days are in the spring and almost contiguous, they illustrate how the DNI can vary considerably from day-to-day. There are 284 instruments recording one-minute data, for a total of 1,244,160 measurements.

This first figure shows the average of all 284 instruments:

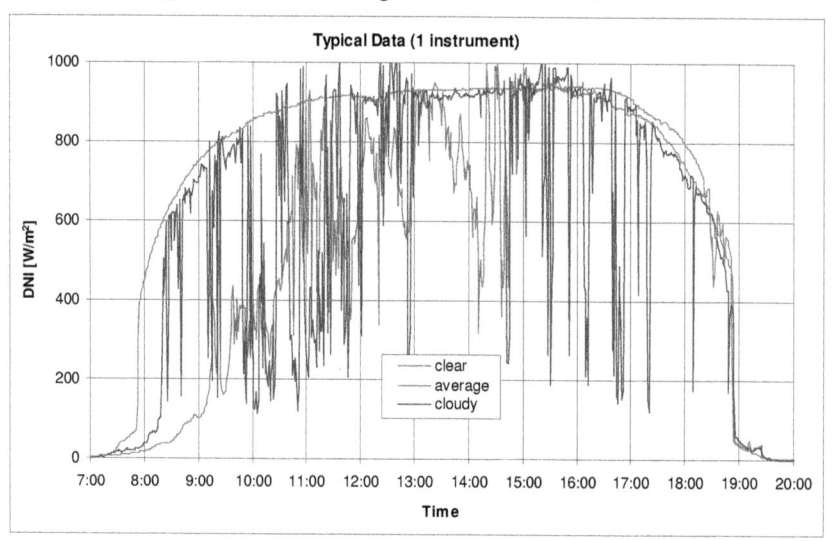

Typical Data (1 instrument)

Typical Data (1 instrument)

This second figure shows typical data from a single instrument. There is considerable variability, even on an average day. This third figure shows the standard deviation of DNI on the clear day.

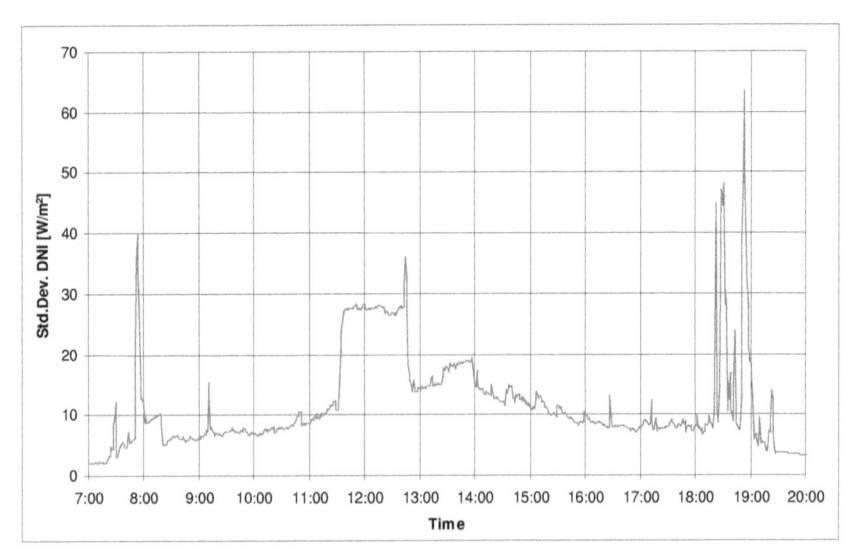

Standard Deviation of DNI on Clear Day

On this clear day the standard deviation approaches 7% of the average. These first figures only show the temporal variability. The spatial variability of DNI at 12:52 on the clear day is 36 W/m^2 and is illustrated in the next figure:

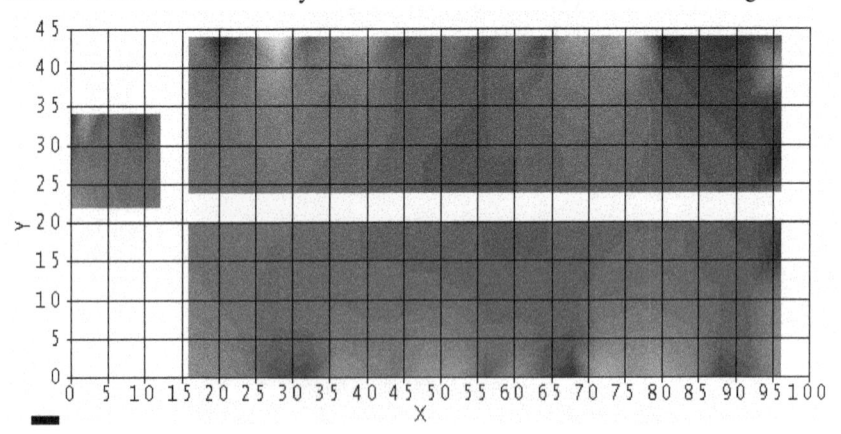

Spatial Variation of DNI at 12:52 on Clear Day

In the above figure the darkest shade corresponds to 909 W/m^2 and the lightest shade corresponds to 975 W/m^2. The standard deviation of DNI at 16:30 on the clear day is 12 W/m^2 and is illustrated in the next figure:

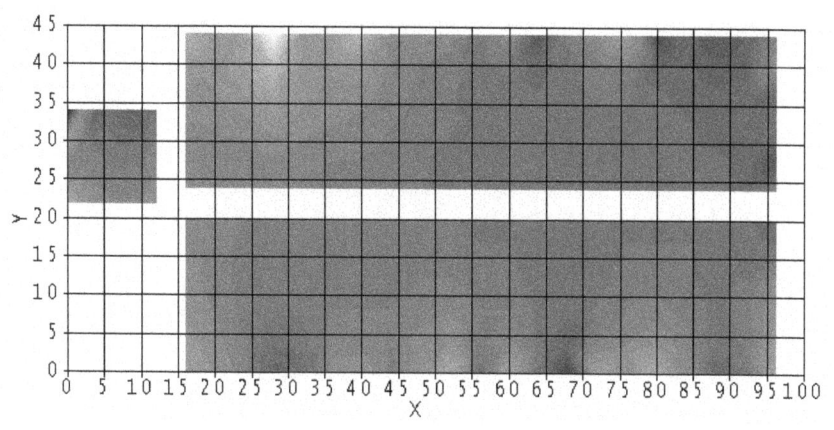

[Chart: Spatial Variation of DNI showing X vs Y axes from 0-100 and 0-45 respectively]

Spatial Variation of DNI at 16:30 on Clear Day

An animation of this data with one frame per minute is included in the archive on the web site. The expected power output of this 500-acre facility for these three days is shown in the next figure.

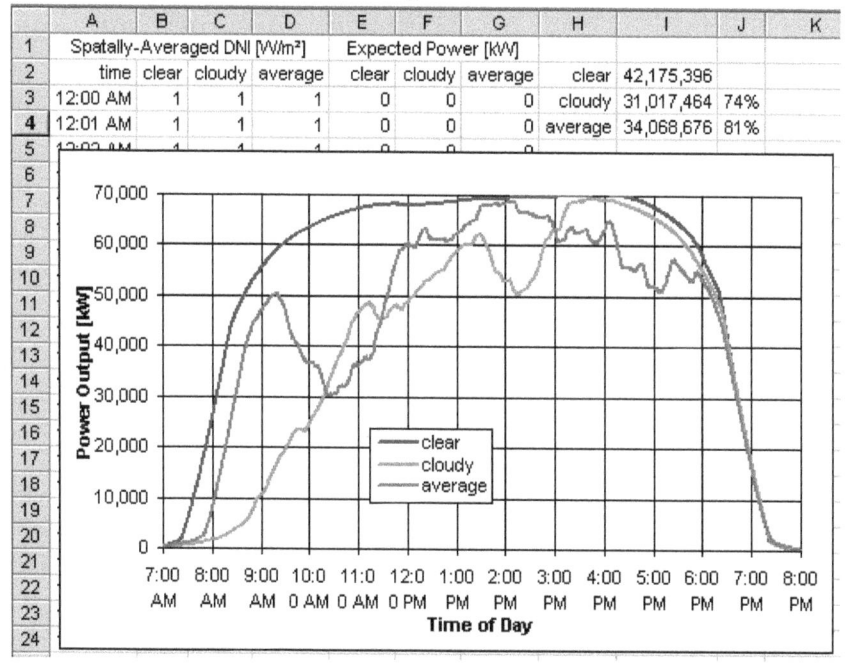

	A	B	C	D	E	F	G	H	I	J	K
1	Spatially-Averaged DNI [W/m²]				Expected Power [kW]						
2	time	clear	cloudy	average	clear	cloudy	average	clear	42,175,396		
3	12:00 AM	1	1	1	0	0	0	cloudy	31,017,464	74%	
4	12:01 AM	1	1	1	0	0	0	average	34,068,676	81%	

[Chart: Power Output [kW] vs Time of Day from 7:00 AM to 8:00 PM, showing clear, cloudy, and average curves]

If you are expecting to recover your investment based on clear-day performance, you will disappointed, as the average is only 81% of this. On a somewhat cloudy day the output is only 74% of that on a clear day. Phoenix,

Arizona is known for it's clear skies, but this is an anecdotal evaluation, not measured data, as in this next figure:

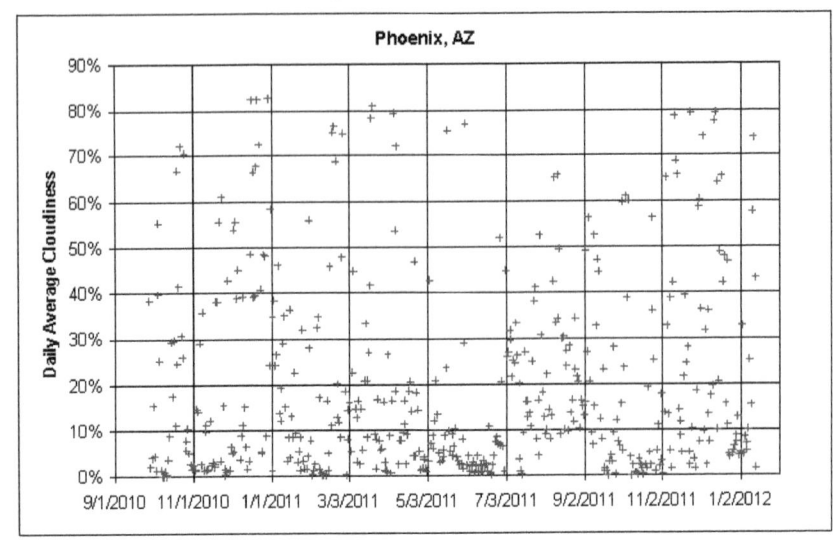

Phoenix, AZ — Daily Average Cloudiness scatter plot from 9/1/2010 to 1/2/2012.

The occurrence of cloudiness is shown in this next figure.

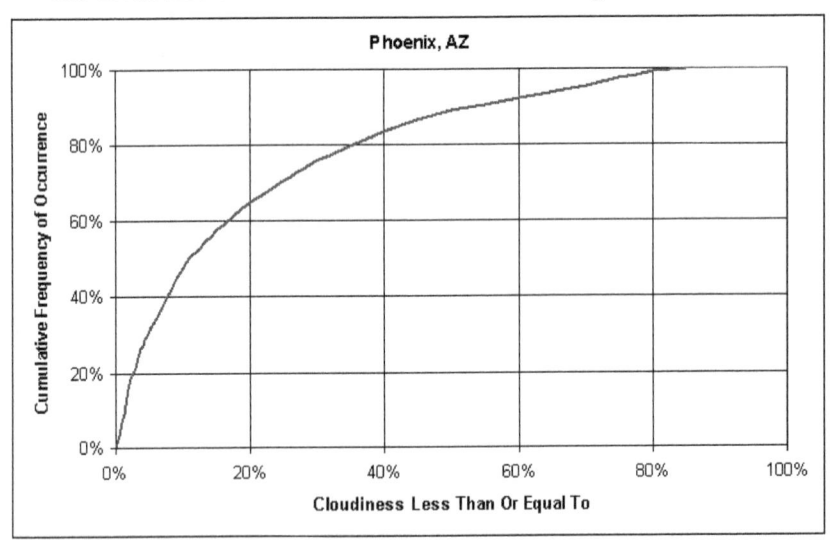

Phoenix, AZ — Cumulative Frequency of Occurrence vs. Cloudiness Less Than Or Equal To.

It is naive to presume clear days, even in Phoenix. You can easily find equations for the position of the sun that account for time of day, day of year, and location on the surface of the Earth. You can also find equations for solar panel output as a function of DNI. What is often lacking is the fraction of the total incident radiation above the atmosphere that will make it to your solar panels. You will need historical data for this. This next figure shows a more reasonable expectation.

Chapter 20. Punkin' Chunkin'

Whether it's a slingshot or a catapult, there's just something wonderful about hurling objects into the sky[8]. The mathematics of projectiles ranges from simple to complex, depending on how fast the projectile is launched and how far it travels. The simplest formulation ignores drag, variations in gravity, and rotation of the Earth. These equations can be found in any introductory text on particle dynamics. Consider a projectile launched at some velocity, V, and angle, θ.

$$x(t) = V \cos(\theta) t$$
$$z(t) = V \sin(\theta) t - \frac{g t^2}{2} \quad (20.1,2)$$

Forward motion stops when $z(t)=0$ and the projectile hits the ground. This occurs at:

$$t = \frac{2 V \sin(\theta)}{g} \quad (20.3)$$

Up until the time of impact, the projectile has traveled (substitute Equation 19.3 into 19.1):

$$x = \frac{2 \sin(\theta) \cos(\theta) V^2}{g} \quad (20.4)$$

Recall the trig identity that $2 \sin(\theta) \cos(\theta) = \sin(2\theta)$, and this becomes:

$$x = \frac{\sin(2\theta) V^2}{g} \quad (20.5)$$

The maximum distance occurs when $dx/d\theta=0$, or:

$$\frac{dx}{d\theta} = 0 = \frac{2 \cos(2\theta) V^2}{g} \quad (20.6)$$

The solution is $\theta=\pi/4$ or 45°, which is consistent with intuition.

---

[8] Little boys never really grow up. Some get a job at NASA so they can hurl really big stuff really far. Others build various devices to hurl pumpkins.

Two common projectile shapes are shown in this next figure:

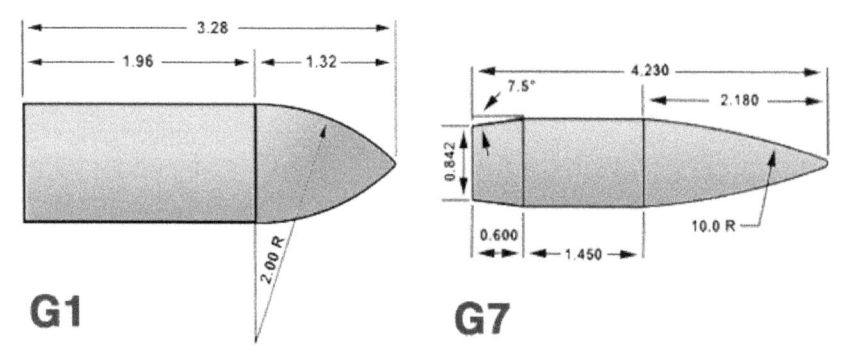

G1　　　　　　　　　　G7

The drag coefficients, C_D, for these two shapes are shown below:

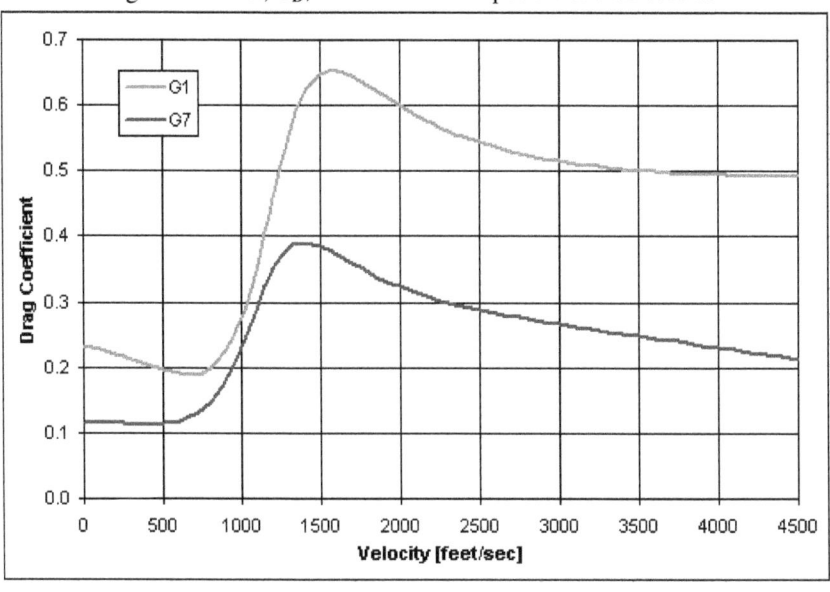

The relationship between drag coefficient and drag force was given in Equation 2.2. The relationships between position, force, and time were given in Equations 1.1 through 1.5. The trajectory of two .30 caliber rounds (having shape G1 and G7, 180 grains each, muzzle velocity of 3000 ft/sec) can be calculated using Euler's method[9]. The results of this simulation are shown in the following figure:

---

[9] Euler's method is the simplest algorithm for solving ordinary differential equations and can be found in any text on numerical methods.

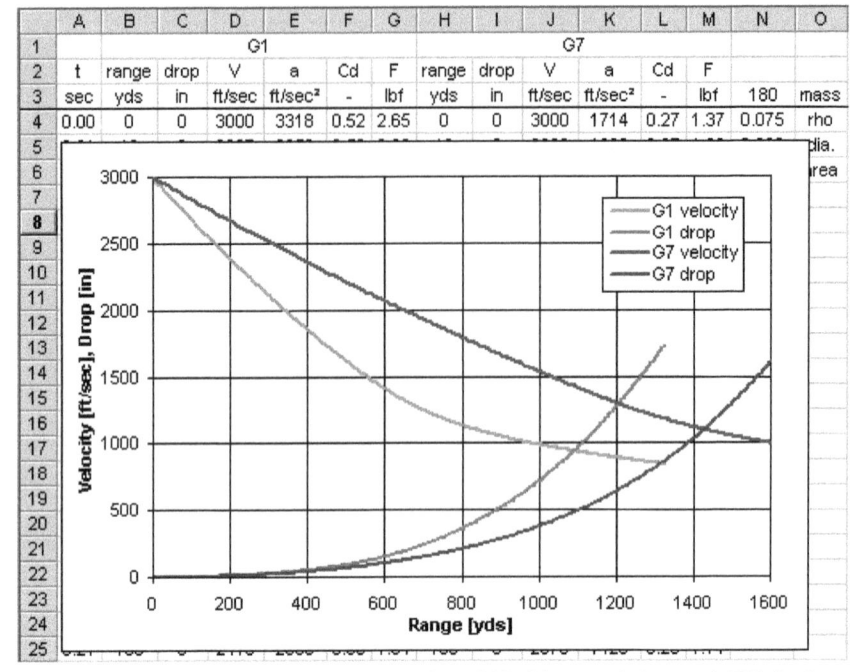

	A	B	C	D	E	F	G	H	I	J	K	L	M	N	O
1				G1						G7					
2	t	range	drop	V	a	Cd	F	range	drop	V	a	Cd	F		
3	sec	yds	in	ft/sec	ft/sec^2	-	lbf	yds	in	ft/sec	ft/sec^2	-	lbf	180	mass
4	0.00	0	0	3000	3318	0.52	2.65	0	0	3000	1714	0.27	1.37	0.075	rho

[Chart showing G1 velocity, G1 drop, G7 velocity, and G7 drop plotted against Range (yds), with Velocity [ft/sec], Drop [in] on the y-axis from 0 to 3000.]

After 1000 yds, the G1 has slowed to 987 ft/sec and dropped 718 inches; whereas, the G7 at this same distance has a velocity of 1533 ft/sec and has only dropped 375 inches. As expected, the G7 shape is a considerable improvement over the G1. Excel® is adequate for this simple problem, but a more robust and versatile solution is preferable as additional complexities are added. For this problem, the most efficient way to solve the differential equations would be 4th order Runge-Kutta[10]. The source code for this is listed in Appendix B.

---

[10] Runge-Kutta methods are often used to solve ordinary differential equations and may be found in any text on numerical methods.

A comparison of results between Euler's method and 4th order Runge-Kutta for the G7 shape is given in this next figure:

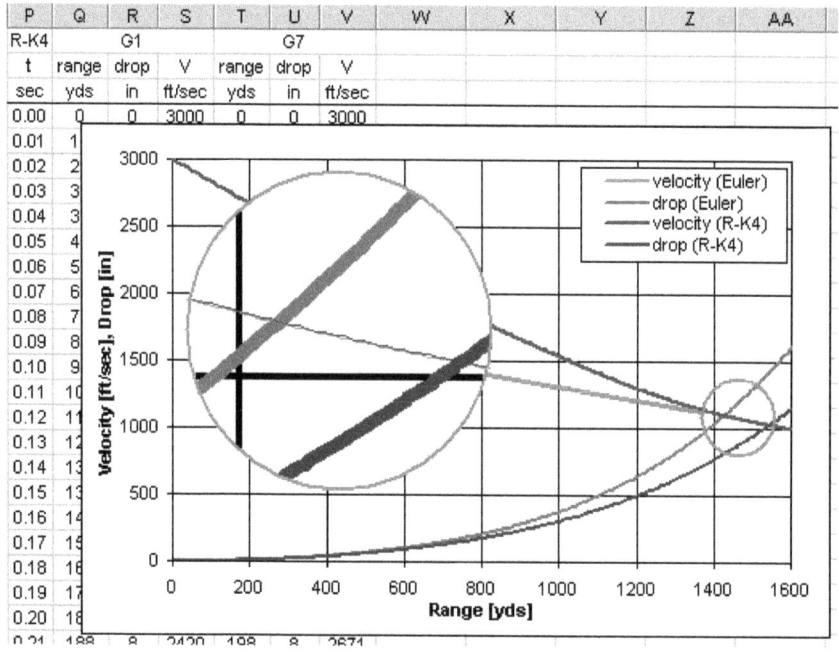

P	Q	R	S	T	U	V	W	X	Y	Z	AA
R-K4		G1			G7						
t	range	drop	V	range	drop	V					
sec	yds	in	ft/sec	yds	in	ft/sec					
0.00	0	0	3000	0	0	3000					
0.01	1										
0.02	2										
0.03	3										
0.04	3										
0.05	4										
0.06	5										
0.07	6										
0.08	7										
0.09	8										
0.10	9										
0.11	10										
0.12	11										
0.13	12										
0.14	13										
0.15	13										
0.16	14										
0.17	15										
0.18	16										
0.19	17										
0.20	18										
0.21	188	8	2430	188	8	2671					

There is hardly any difference in velocity, but a noticeable difference in drop. This is because in the Excel® spreadsheet (i.e., Euler's method), we ignored the fact that the drag force is in line with the velocity, which is not exactly horizontal. We can now use this same program to determine the azimuth resulting in the maximum range.

The results are illustrated in this next figure:

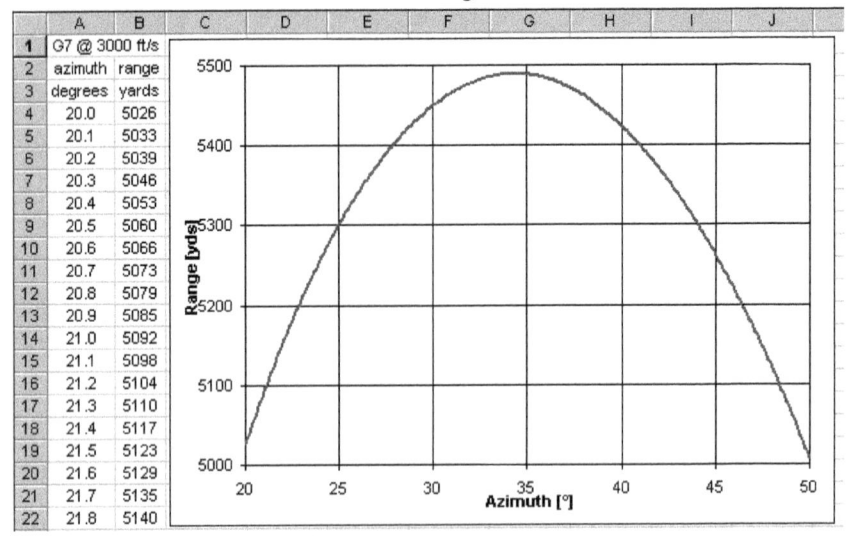

	A	B
1	G7 @ 3000 ft/s	
2	azimuth	range
3	degrees	yards
4	20.0	5026
5	20.1	5033
6	20.2	5039
7	20.3	5046
8	20.4	5053
9	20.5	5060
10	20.6	5066
11	20.7	5073
12	20.8	5079
13	20.9	5085
14	21.0	5092
15	21.1	5098
16	21.2	5104
17	21.3	5110
18	21.4	5117
19	21.5	5123
20	21.6	5129
21	21.7	5135
22	21.8	5140

The maximum range is 5489 yards, which occurs at an azimuth of 34.4°, considerably different from the ideal value of 45°.

We will next consider a punkin' chunkin' air cannon similar to the one pictured below:

The overall length of the cannon is 120 ft with a barrel length of 90 ft. The pumpkin weighs about 10 pounds and the range is approximately 4700 ft.

We will use the drag coefficient for a smooth sphere for our simulation, as given by the following figure:

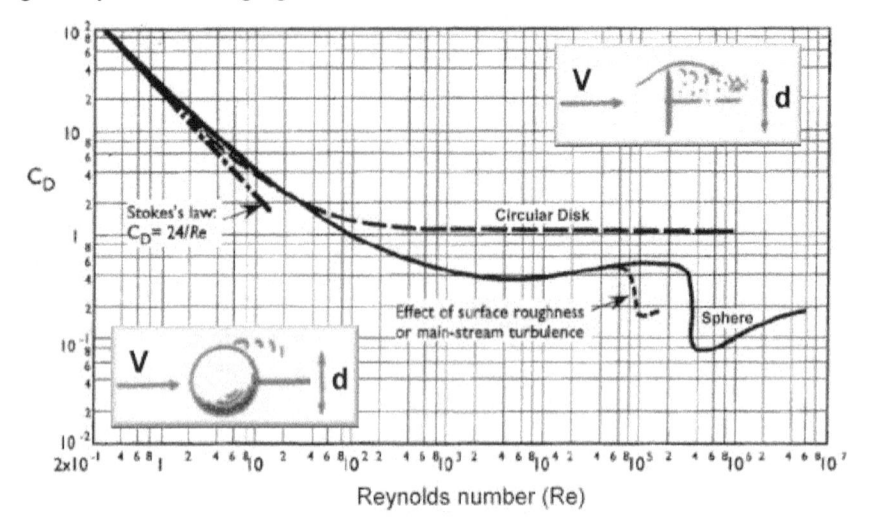

With an initial velocity of 600 ft/sec and an azimuth of 45°, we obtain the following trajectory:

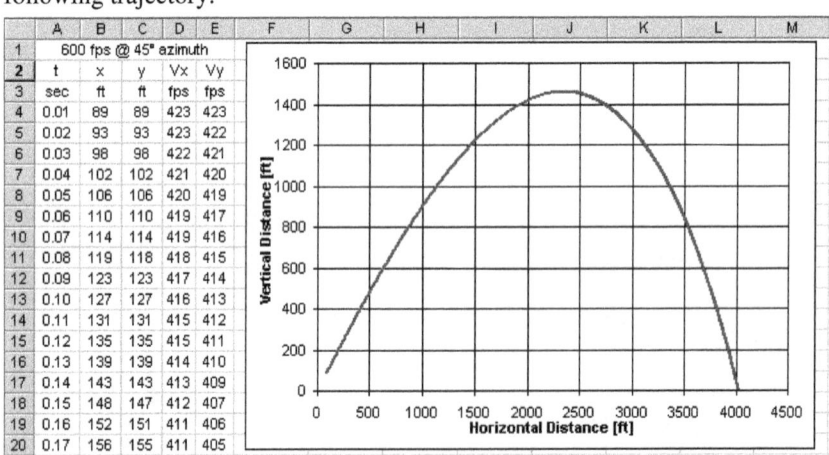

	A	B	C	D	E
1	600 fps @ 45° azimuth				
2	t	x	y	Vx	Vy
3	sec	ft	ft	fps	fps
4	0.01	89	89	423	423
5	0.02	93	93	423	422
6	0.03	98	98	422	421
7	0.04	102	102	421	420
8	0.05	106	106	420	419
9	0.06	110	110	419	417
10	0.07	114	114	419	416
11	0.08	119	118	418	415
12	0.09	123	123	417	414
13	0.10	127	127	416	413
14	0.11	131	131	415	412
15	0.12	135	135	415	411
16	0.13	139	139	414	410
17	0.14	143	143	413	409
18	0.15	148	147	412	407
19	0.16	152	151	411	406
20	0.17	156	155	411	405

This time the maximum range is 4091 ft and is obtained with an azimuth of 39°. As this example shows, it's not obvious what the optimum azimuth will be. It depends on the shape and initial velocity. It can also be influenced by any wind present.

Punkin' chunkin' is a science!

The program to calculate the punkin' trajectory is listed in Appendix C. We can modify this program to account for cross wind, by adding another dimension. This is a definite advantage to using the Runge-Kutta method for solving the differential equations. The following curve shows the impact of a 20 mph cross wind.

	A	B	C	D	E	F	G
1	20 mph cross wind - azimuth 39°						
2	t	x	y	z	Vx	Vy	Vz
3	sec	ft	ft	ft	fps	fps	fps
4	0.00	93	0	76	466	0	378
5	0.01	98	0	79	465	0	377
6	0.02	103	0	83	464	0	375
7	0.03	107	0	87	464	0	374
8	0.04	112	0	91	463	0	373
9	0.05	116	0	94	462	0	372
10	0.06	121	0	98	461	0	371
11	0.07	126	0	102	460	0	370
12	0.08	130	0	105	459	0	369
13	0.09	135	0	109	458	1	368
14	0.10	139	0	113	457	1	367
15	0.11	144	0	116	456	1	366
16	0.12	149	0	120	456	1	365
17	0.13	153	0	124	455	1	364
18	0.14	158	0	127	454	1	363
19	0.15	162	0	131	453	1	362
20	0.16	167	0	135	452	1	361

After 4088 feet of travel in the direction of the launch, the punkin' has been blown sideways by 242 feet.

This next figure shows the impact for a range of cross winds:

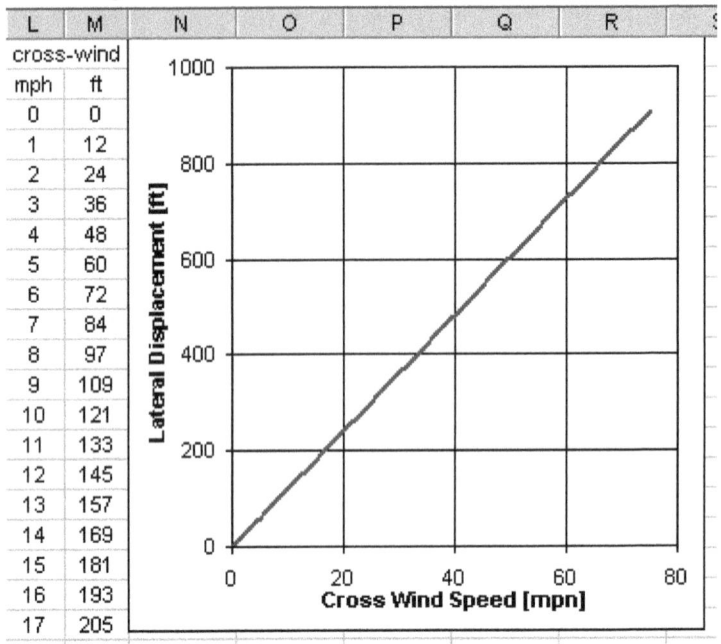

cross-wind	
mph	ft
0	0
1	12
2	24
3	36
4	48
5	60
6	72
7	84
8	97
9	109
10	121
11	133
12	145
13	157
14	169
15	181
16	193
17	205

We can use this same 4th order Runge-Kutta method for calculating what happens to the punkin' in the barrel when the valve is opened. The air tank used in this cannon was originally built to hold 1000 gallons of propane. The pressure will drop as it expands into the barrel and begins pushing the punkin'. It would be too optimistic to assume the temperature of the air will remain constant during this process. The best scenario would be an isentropic process, in which case the air would expand according to the following relationship:

$$\left(\frac{P_2}{P_1}\right) = \left(\frac{V_1}{V_2}\right)^k$$

$$k = \frac{C_P}{C_v}$$

(20.6,7)

In this equation k is called the isentropic exponent and is equal to the ratio of the constant pressure and constant volume specific heats. For air k is equal to 1.4. The volume will increase by the distance the punkin' is displaced times the area of the barrel. We must also expect some pressure drop through the valve and piping. The results are shown in the following figure:

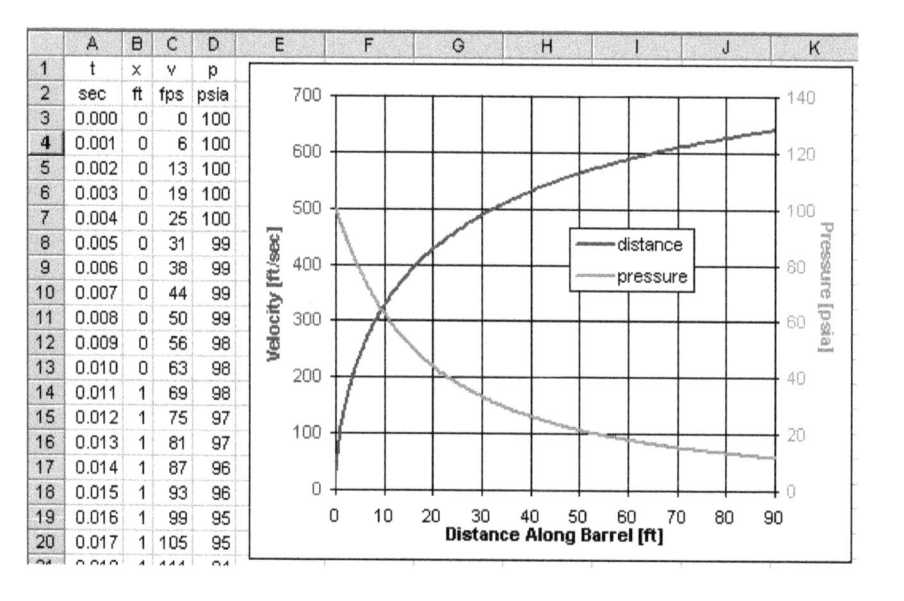

	A	B	C	D
1	t	x	v	p
2	sec	ft	fps	psia
3	0.000	0	0	100
4	0.001	0	6	100
5	0.002	0	13	100
6	0.003	0	19	100
7	0.004	0	25	100
8	0.005	0	31	99
9	0.006	0	38	99
10	0.007	0	44	99
11	0.008	0	50	99
12	0.009	0	56	98
13	0.010	0	63	98
14	0.011	1	69	98
15	0.012	1	75	97
16	0.013	1	81	97
17	0.014	1	87	96
18	0.015	1	93	96
19	0.016	1	99	95
20	0.017	1	105	95

Chapter 21. Bungee Cords

Bungee cords are made of rubber and are typically 5/8th inch in diameter and perhaps 30 feet long. As a force is applied to the cord, it stretches and absorbs energy until the point at which it fails. The point of failure for the type of rubber most often used is about 1850 pounds per square inch (psi). This failure occurs at an elongation of approximately 182%. This next figure illustrates the relationship between the stress (force per unit area), strain (percent elongation), and stored energy:

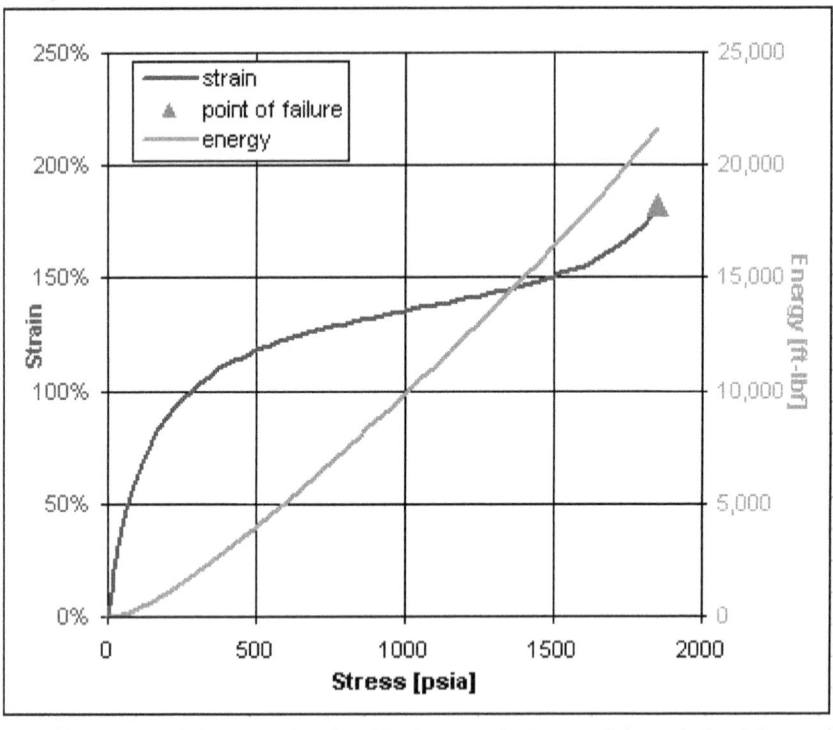

The slope of the stress/strain (blue) curve is the modulus of elasticity and can be approximated by the following formula:

$$\varepsilon = \frac{1.42374\,\sigma\,(2187.54 - \sigma)}{(\sigma + 137.568)(2095.8 - \sigma)} \qquad (21.1)$$

In this case σ is the stress and ε is the strain. The stored energy, v, is found by integrating the strain with respect to the stress, or:

$$v = \int_0^\sigma \varepsilon\, d\sigma \qquad (21.2)$$

Equation 20.1 can be integrated analytically to obtain:

$$v = \begin{cases} 1941.7 + 1.4237\sigma \\ -203.91\ln(\sigma + 137.568) \\ -122.6\ln(2095.8 - \sigma) \end{cases} \quad (21.3)$$

The initial cross-sectional area of the 5/8ths inch cord is 1.227 in² so that a 250-pound load would result in a stress of 203.7 psi. This would result in a strain of 89.1% or an elongated length of 56.7 ft. The stored energy would be 1080 foot-pounds (ft-lbf). The loading curve for this cord would look just like the stress/strain curve, but with a different scale (i.e., stress*1.227 in² and (strain+1)*30 ft).

If we were to dead drop a 250-pound weight (Bubba) onto this cord, it would free fall for 30 feet before the cord would begin to stretch and absorb the energy. The downward motion would stop at the point where the gravitational energy of the falling weight is equal to the energy absorbed by the cord, as the velocity would then be zero. This distance is found by solving the following equation:

$$v(h - 30) = mgh \quad (21.4)$$

Equation 20.4 is illustrated graphically in the following figure:

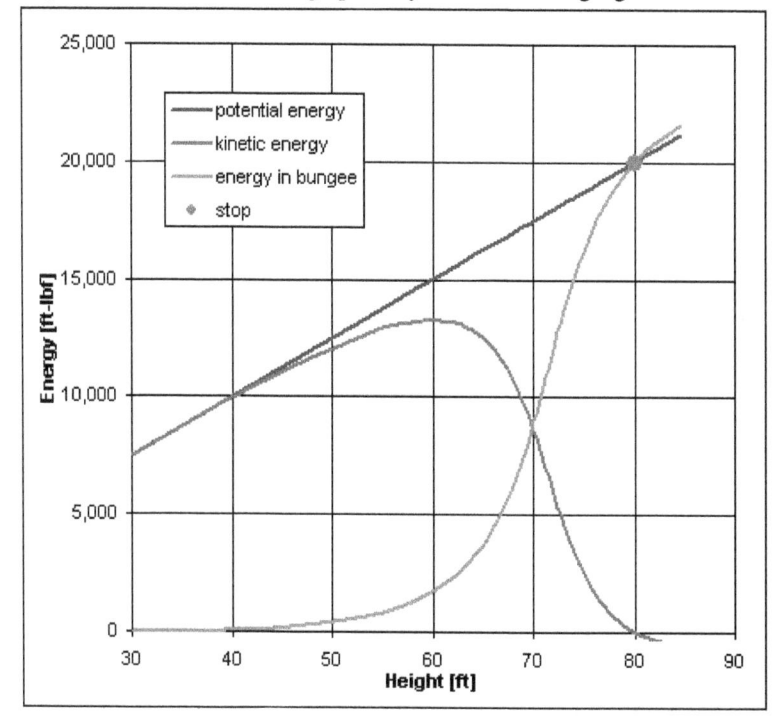

107

The straight blue line represents the potential energy. The green curve is the kinetic energy, which rises as Bubba accelerates downward and then falls as the cord pulls him to a stop. The orange curve represents the energy stored in the bungee. The red dot is where the blue and orange curves intersect and the green curve reaches zero. This occurs at 80 ft and 20,000 ft-lbf. There's a function in the spreadsheet to solve this equation for you.

The red triangle in the previous figure showed the failure point for this material, which occurs at 21,650 ft-lbf for this length and diameter cord. This means that Bubba came mighty close to breaking the cord. We can use this same calculation to find the drop length of this bungee for a range of weights, as shown in this next figure:

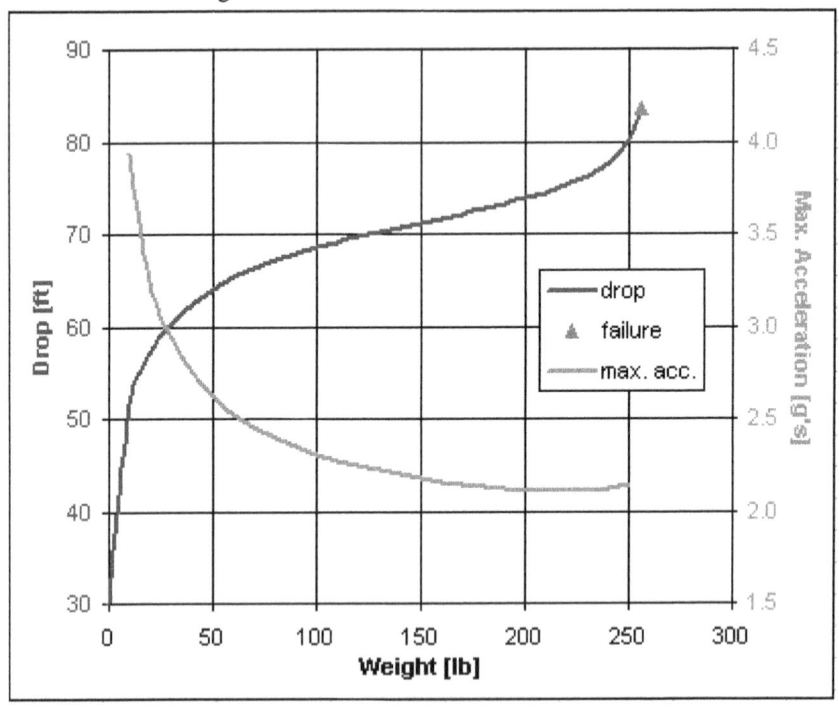

The failure point for this cord is 255.5 pounds. The maximum acceleration occurs at the bottom of the drop and is given by Equation 1.5. The force and mass are known, so the acceleration is easily calculated and is shown on the figure above as the orange curve.

Appendix A. Engine Clearance Simulation Program

This is the code used for the Monte Carlo simulation of engine clearance:

```
#include <stdio.h>
#include <stdlib.h>
#include <math.h>

int urand()
  {
  int i,u;
  for(u=i=0;i<12;i++)
    u+=rand();
  return(u);
  }

double rnorm()
  {
  return(urand()/32767./6.-1.);
  }

double rdist(double a,double s)
  {
  return(a+6.*s*rnorm());
  }

int nint(double d)
  {
  if(d>0.)
    return((int)(d+0.5));
  if(d<0.)
    return((int)(d-0.5));
  return(0);
  }

#define cylinder()  rdist(3.73820,0.00061)
#define   piston()  rdist(3.73600,0.00058)

int engine1[7];     /* engines out of spec 1 */
int engine2[7];     /* engines out of spec 2 */
int count[2000];    /* clearance bins */
double clear[6];    /* 6-cylinder clearances */
double miles[6];    /* 6-cylinder longevities */

void clearance()
  {
  int i,j,k;
  double c,p;
  for(i=0;i<6000006;i++)
    {
```

```
    c=cylinder();
    p=piston();
    j=nint((c-p)*100000.)+250;
    count[j]++;
    }
  for(i=0;i<sizeof(count)/sizeof(count[0]);i++)
    if(count[i])
      printf("%7.5lf\t%i\n",(i-250)/100000.,count[i]);
  }

void engines()
  {
  int i,j,k;
  double c,p;
  for(i=0;i<1000006;i++)
    {
    c=cylinder();
    p=piston();
    j=nint((c-p)*100000.)+250;
    clear[i%6]=c-p;
    if(i>=6)
      {
      for(j=k=0;j<6;j++)
        if(clear[j]<0.00140||clear[j]>0.00300)
          k++;
      engine1[k]++;
      for(j=k=0;j<6;j++)
        if(clear[j]<0.00080||clear[j]>0.00360)
          k++;
      engine2[k]++;
      }
    }
  for(i=0;i<=6;i++)
    printf("%i\t%i\t%i\n",i,engine1[i],engine2[i]);
  }

double life(double clear)
  {
  if(clear<=0.)
    return(0.);
  if(clear<0.00220)
    return(160000.*pow(clear/0.00220,2));
  if(clear>0.00220)
    return(160000.*0.00220/clear);
  return(160000.);
  }

void longevity()
  {
```

```
  int i,j;
  double c,m,p;
  memset(count,0,sizeof(count));
  for(i=0;i<1000006;i++)
    {
    c=cylinder();
    p=piston();
    miles[i%6]=life(c-p);
    if(i>=6)
      {
      m=miles[0];
      for(j=1;j<6;j++)
        if(miles[j]<m)
          m=miles[j];
      j=nint(m/5000.);
      count[j]++;
      }
    }
  for(i=0;i<sizeof(count)/sizeof(count[0]);i++)
    if(count[i])
      printf("%i\t%i\n",5000*i,count[i]);
  }

int main(int argc,char**argv,char**envp)
  {
  clearance();
  engines();
  longevity();
  return(0);
  }
```

Appendix B. Projectile Trajectory Using 4th Order Runge-Kutta

The following code uses the 4th order Runge-Kutta method to solve for the trajectory of a projectile:

```
#define _CRT_SECURE_NO_DEPRECATE
#include <stdio.h>
#include <stdlib.h>
#include <float.h>
#define _USE_MATH_DEFINES
#include <math.h>

void rk4(void
   dYdX(double,double*,double*),double*X,double
   dX,double*Y,double*dY,int n)
{/* 4th-Order Runge-Kutta */
int i,j;
static double
   A[3]={0.5,0.5,1.},B[4]={1./6.,1./3.,1./3.,1./6.};
double*V,*W,Xi;
V=calloc(5*n,sizeof(double));
W=V+n;
Xi=*X;
dYdX(Xi,Y,W);
for(i=1;i<4;i++)
   {
   Xi=*X+dX*A[i-1];
   for(j=0;j<n;j++)
      {
      dY[j]=A[i-1]*W[n*(i-1)+j];
      V[j]=Y[j]+dX*dY[j];
      }
   dYdX(Xi,V,W+n*i);
   }
for(j=0;j<n;j++)
   {
   dY[j]=0.;
   for(i=0;i<4;i++)
      dY[j]+=B[i]*W[n*i+j];
   Y[j]+=dX*dY[j];
   }
*X+=dX;
free(V);
}

double G7(double V)
   {
   return(((1.17113760313435E-7*V-2.08047377740163E-
   4)*V+0.11589995765079)/
```

```
        ((((2.13378583637085E-14*V-1.58952440981953E-
      10)*V+9.79077267524975E-7)
      *V-1.73323825318021E-3)*V+1.));
    }

    double G1(double V)
      {
      return(((2.21164801772262E-7*V-4.17387420148189E-
      4)*V+0.234113648441904)/
      (((-2.04443572897802E-11*V+6.6060544141452E-7)*V-
      1.50268271146268E-3)*V+1.));
      }

    #define diameter  0.3   /* diameter [inch] */
    #define grains    180.  /* mass [grains] */
    #define Vmuzzle   3000. /* muzzle velocity [ft/sec] */
    #define height    6.    /* height of discharge [ft] */
    #define azimuth   0.    /* angle with respect to
       horizontal [degrees] */
    #define rho   0.074887  /* density of air at 70°F
       [lbm/ft³] */
    #define g         32.174 /* acceleration of gravity
       [ft/sec²] */

    #define area (M_PI*(diameter/12.)*(diameter/12.)/4.)
    #define mass (grains/437.5/16.)

    void bullet(double t,double*B,double*dB)
      {
      double accel,angle,Cd,Fdrag,V,Vx,Vy,X,Y;
      X=B[0];
      Y=B[1];
      Vx=B[2];
      Vy=B[3];
      angle=atan2(Vy,Vx);
      V=hypot(Vx,Vy);
      Cd=G7(V);
      Fdrag=area*Cd*rho*V*V/2./g;
      accel=-g*Fdrag/mass;
      dB[0]=Vx;
      dB[1]=Vy;
      dB[2]=accel*cos(angle);
      dB[3]=accel*sin(angle)-g;
      }

    int main(int argc,char**argv,char**envp)
      {
      double t,B[4],dB[4];
      B[0]=B[1]=B[3]=0.;
```

```c
  B[2]=Vmuzzle;
  printf("t\trange\tdrop\tV\n");
  printf("sec\tyds\tin\tft/sec\n");
  for(t=0.;t<3.;)
    {
    rk4(bullet,&t,0.01,B,dB,4);
    printf("%.2lf\t%.0lf\t%.0lf\t%.0lf\n",t,B[0]/3.,-
  12.*B[1],hypot(B[2],B[3]));
    }
  return(0);
  }
```

Appendix C. Punkin' Trajectory and Propulsion

The following code uses the 4th order Runge-Kutta method to solve for the trajectory of a punkin and also the propulsion through the barrel when the valve is opened:

```
/* Punkin' Chunkin' Trajectory using 4th Order Runge-
   Kutta
   position is in modified spherical coordinates.
   position
     X=B[0] horizontal in the direction of launch
     Y=B[1] horizontal in the direction of cross wind
     Z=B[2] vertical
   velocities
     Vx=dX/dt=dB[0]=B[3]
     Vy=dY/dt=dB[1]=B[4]
     Vz=dZ/dt=dB[2]=B[5]
   acceleration
     Ax=dVx/dt=dB[3]
     Ay=dVy/dt=dB[4]
     Az=dVz/dt=dB[5]
   differential equation
     Fx=m*Ax=Fdrag*cos(è)*cos(é)
     Fy=m*Ay=Fdrag*sin(è)
     Fz=m*Az=Fdrag*cos(è)*sin(é)-g
**********************************/

#define _CRT_SECURE_NO_DEPRECATE
#include <stdio.h>
#include <stdlib.h>
#include <float.h>
#define _USE_MATH_DEFINES
#include <math.h>

double hypot3(double x,double y,double z)
  {
    return(sqrt(x*x+y*y+z*z));
  }

void rk4(void
    dYdX(double,double*,double*),double*X,double
    dX,double*Y,double*dY,int n)
  {/* 4th-Order Runge-Kutta */
   int i,j;
   static double
     A[3]={0.5,0.5,1.},B[4]={1./6.,1./3.,1./3.,1./6.};
   double*V,*W,Xi;
   V=calloc(5*n,sizeof(double));
   W=V+n;
   Xi=*X;
```

```
  dYdX(Xi,Y,W);
  for(i=1;i<4;i++)
    {
    Xi=*X+dX*A[i-1];
    for(j=0;j<n;j++)
      {
      dY[j]=A[i-1]*W[n*(i-1)+j];
      V[j]=Y[j]+dX*dY[j];
      }
    dYdX(Xi,V,W+n*i);
    }
  for(j=0;j<n;j++)
    {
    dY[j]=0.;
    for(i=0;i<4;i++)
      dY[j]+=B[i]*W[n*i+j];
    Y[j]+=dX*dY[j];
    }
  *X+=dX;
  free(V);
  }

struct{double Re,Cd;}cdata[]={{-0.13, 1.53},{ 0.02,
  1.40},{ 0.17, 1.27},
  { 0.32, 1.13},{ 0.46, 0.99},{ 0.62, 0.87},{ 0.78,
  0.74},{ 0.94, 0.63},
  { 1.11, 0.51},{ 1.27, 0.40},{ 1.45, 0.30},{ 1.62,
  0.21},{ 1.80, 0.11},
  { 1.98, 0.02},{ 2.16,-0.06},{ 2.35,-0.13},{ 2.54,-
  0.20},{ 2.73,-0.26},
  { 2.92,-0.32},{ 3.11,-0.36},{ 3.31,-0.39},{ 3.51,-
  0.41},{ 3.71,-0.42},
  { 3.91,-0.41},{ 4.11,-0.39},{ 4.31,-0.36},{ 4.51,-
  0.33},{ 4.71,-0.30},
  { 4.91,-0.29},{ 5.11,-0.27},{ 5.31,-0.29},{ 5.47,-
  0.39},{ 5.53,-0.58},
  { 5.56,-0.78},{ 5.62,-0.97},{ 5.79,-1.06},{ 5.98,-
  1.00},{ 6.17,-0.92},
  { 6.35,-0.84},{ 6.54,-0.78},{ 6.74,-0.74},{ 6.94,-
  0.73},{ 7.14,-0.72}};

double Cd(double Re)
  {
  int i;
  double logRe;
  logRe=log10(Re);
  for(i=1;i<sizeof(cdata)/sizeof(cdata[0]);i++)
    if(logRe>=cdata[i-1].Re)
      if(logRe<=cdata[i].Re)
```

```
        return(pow(10.,cdata[i-1].Cd+(cdata[i].Cd-
    cdata[i-1].Cd)*(logRe-cdata[i-1].Re)/(cdata[i].Re-
    cdata[i-1].Re)));
    if(Re<2.)
        return(24./Re);
    return(pow(10.,-0.72));
}

#define Length    120.    /* overall length [ft] */
#define diameter  10.     /* diameter [inch] */
#define mass      10.     /* mass [pounds] */
#define Vmuzzle   600.    /* muzzle velocity [ft/sec] */
#define rho       0.0754  /* density of air at 70°F
    [lbm/ft³] */
#define mu        0.0443  /* dynamic viscosity of air
    [lbm/ft/hr] */
#define g         32.174  /* acceleration of gravity
    [ft/sec²] */
double wind     = 20.;  /* cross wind [mph] */

#define area (M_PI*(diameter/12.)*(diameter/12.)/4.)

void trajectory(double t,double*B,double*dB)
{
    double accel,Fdrag,phi,Re,theta,V,Vw,Vx,Vy,Vz;
    Vx=B[3];
    Vy=B[4];
    Vz=B[5];
    Vw=wind*5280./3600.;
    V=hypot3(Vx,Vy-Vw,Vz);
    theta=atan2(Vz,Vx);
    phi=asin((Vy-Vw)/V);
    Re=(diameter/12.)*rho*V/(mu/3600.);
    Fdrag=area*Cd(Re)*rho*V*V/2./g;
    accel=-g*Fdrag/mass;
    dB[0]=Vx;
    dB[1]=Vy;
    dB[2]=Vz;
    dB[3]=accel*cos(phi)*cos(theta);
    dB[4]=accel*sin(phi);
    dB[5]=accel*cos(phi)*sin(theta)-g;
}

double range(double azimuth)
{
    double B[6],dB[6],t,X1,X2,Y1,Y2,Z1,Z2;
    B[0]=Length*cos(M_PI*azimuth/180.);
    B[1]=B[4]=0.;
    B[2]=Length*sin(M_PI*azimuth/180.);
```

```
    B[3]=Vmuzzle*cos(M_PI*azimuth/180.);
    B[5]=Vmuzzle*sin(M_PI*azimuth/180.);
//printf("t\tx\ty\tz\tVx\tVy\tVz\n");
//printf("sec\tft\tft\tfps\tfps\n");
    X2=Y2=Z2=t=0.;
//printf("%.2lf\t%.0lf\t%.0lf\t%.0lf\t%.0lf\t%.0lf\t%.0l
    f\n",t,B[0],B[1],B[2],B[3],B[4],B[5]);
    do{
        rk4(trajectory,&t,0.01,B,dB,6);
//
        printf("%.2lf\t%.0lf\t%.0lf\t%.0lf\t%.0lf\t%.0lf\t%.0
        lf\n",t,B[0],B[1],B[2],B[3],B[4],B[5]);
        X1=X2;
        Y1=Y2;
        Z1=Z2;
        X2=B[0];
        Y2=B[1];
        Z2=B[2];
        }while(B[2]>=0.);
//return(X1-(X2-X1)*Z1/(Z2-Z1));
    return(Y1-(Y2-Y1)*Z1/(Z2-Z1));
    }

#define k       1.40 /* isentropic exponent [unitless] */
#define Vo      100. /* initial volume [gallons] */
#define Po      100. /* initial pressure [psia] */
#define barrel  90.  /* length of barrel [ft] */
#define dP      0.75 /* fractional pressure drop through
    valve */

void propulsion(double t,double*B,double*dB)
    {
    double F,P,V;
    V=Vo+area*B[0]*7.480519481;
    P=Po*pow(Vo/V,k);
    F=144.*P*(1.-dP)*area;
    dB[1]=g*F/mass;
    dB[0]=B[1];
    B[2]=P;
    }

void launch()
    {
    double B[3],dB[2],t;
    B[0]=B[1]=0.;
    B[2]=Po;
    printf("t\tx\tv\tp\n");
    printf("sec\tft\tfps\tpsia\n");
    t=0.;
```

```
      printf("%.3lf\t%.2lf\t%.2lf\t%.2lf\n",t,B[0],B[1],B[2
      ]);
   do{
      rk4(propulsion,&t,0.001,B,dB,2);

      printf("%.3lf\t%.2lf\t%.2lf\t%.2lf\n",t,B[0],B[1],B[2
      ]);
      }while(B[0]<=barrel);
   }

int main(int argc,char**argv,char**envp)
   {
   //double azimuth=39.;
   //for(azimuth=20.;azimuth<=60.;azimuth+=1.)
   //    printf("%lG\t%lG\n",azimuth,range(azimuth));
   //for(wind=0.;wind<75.1;wind+=1.)
   //    printf("%lG\t%lG\n",wind,range(azimuth));
      launch();
      return(0);
   }
```

119

also by D. James Benton

3D Rendering in Windows: How to display three-dimensional objects in Windows with and without OpenGL, ISBN-9781520339610, Amazon, 2016.

A Synergy of Short Stories: The whole may be greater than the sum of the parts, ISBN-9781520340319, Amazon, 2016.

Azeotropes: Behavior and Application, ISBN-9798609748997, Amazon, 2020.

bat-Elohim: Book 3 in the Little Star Trilogy, ISBN-9781686148682, Amazon, 2019.

Complex Variables: Practical Applications, ISBN-9781794250437, Amazon, 2019.

Compression & Encryption: Algorithms & Software, ISBN-9781081008826, Amazon, 2019.

Computational Fluid Dynamics: an Overview of Methods, ISBN-9781672393775, Amazon, 2019.

Computer Simulation of Power Systems: Programming Strategies and Practical Examples, ISBN-9781696218184, Amazon, 2019.

Curve-Fitting: The Science and Art of Approximation, ISBN-9781520339542, Amazon, 2016.

Differential Equations: Numerical Methods for Solving, ISBN-9781983004162, Amazon, 2018.

Evaporative Cooling: The Science of Beating the Heat, ISBN-9781520913346, Amazon, 2017.

Heat Exchangers: Performance Prediction & Evaluation, ISBN-9781973589327, Amazon, 2017.

Heat Recovery Steam Generators: Thermal Design and Testing, ISBN-9781691029365, Amazon, 2019.

Jamie2 2nd Ed.: Innocence is easily lost and cannot be restored, ISBN-9781520339375, Amazon, 2016-18.

Little Star 2nd Ed.: God doesn't do things the way we expect Him to. He's better than that! ISBN-9781520338903, Amazon, 2015-17.

Living Math: Seeing mathematics in every day life (and appreciating it more too), ISBN-9781520336992, Amazon, 2016.

Lost Cause: If only history could be changed…, ISBN-9781521173770. Amazon, 2017.

Mill Town Destiny: The Hand of Providence brought them together to rescue the mill, the town, and each other, ISBN-9781520864679, Amazon, 2017.

Monte Carlo Simulation: The Art of Random Process Characterization, ISBN-9781980577874, Amazon, 2018.

Nonlinear Equations: Numerical Methods for Solving, ISBN-9781717767318, Amazon, 2018.

Numerical Calculus: Differentiation and Integration, ISBN-9781980680901, Amazon, 2018.

Particle Tracking: Computational Strategies and Diverse Examples, ISBN-9781692512651, Amazon, 2019.

Plumes: Delineation & Transport, ISBN-9781702292771, Amazon, 2019.

Power Plant Performance Curves: for Testing and Dispatch, ISBN-9798640192698, Amazon, 2020.

ROFL: Rolling on the Floor Laughing, ISBN-9781973300007, Amazon, 2017.

Steam 2020: to 150 GPa and 6000 K, ISBN-9798634643830, Amazon, 2020.

The Last Seraph: Sequel to *Little Star*, ISBN-9781726802253, Amazon, 2018.

The Many Uses of Orthogonal Functions, ISBN-9781719876162, Amazon, 2018.

The Waterfront Murders: As you sow, so shall you reap, ISBN-9798611314500, Amazon, 2020.

Thermochemical Reactions: Numerical Solutions, ISBN-9781073417872, Amazon, 2019.

Thermodynamic and Transport Properties of Fluids, ISBN-9781092120845, Amazon, 2019.

Thermodynamic Cycles: Effective Modeling Strategies for Software Development, ISBN-9781070934372, Amazon, 2019.

Thermodynamics - Theory & Practice: The science of energy and power, ISBN-9781520339795, Amazon, 2016.

Version-Independent Programming: Code Development Guidelines for the Windows® Operating System, ISBN-9781520339146, Amazon, 2016.

Lightning Source UK Ltd.
Milton Keynes UK
UKHW011905310121
378006UK00004B/34